CITY OF FORESTS, CITY OF FARMS

CITY OF FORESTS, CITY OF FARMS

Sustainability Planning for New York City's Nature

Lindsay K. Campbell

CORNELL UNIVERSITY PRESS ITHACA AND LONDON

The U.S. Department of Agriculture, Forest Service provided funding support for this research.

First published 2017 by Cornell University Press

Printed in the United States of America

Library of Congress Cataloging-in-Publication Data

Names: Campbell, Lindsay K.
Title: City of forests, city of farms : sustainability planning for New York City's nature / Lindsay K. Campbell.
Description: Ithaca : Cornell University Press, 2017. | Includes bibliographical references and index.
Identifiers: LCCN 2017016806 (print) | LCCN 2017020164 (ebook) | ISBN 9781501714702 (epub/mobi) | ISBN 9781501714795 (pdf) | ISBN 9781501707506 (cloth : alk. paper)
Subjects: LCSH: Human ecology—New York (State)—New York. | Urban forestry—New York (State)—New York. | Urban agriculture—New York (State)—New York. | Urban ecology (Biology)—New York (State)—New York. | Green movement—New York (State)—New York. | Sustainable living—New York (State)—New York. | Land use, Urban—Environmental aspects—New York (State)—New York. | City planning—Environmental aspects—New York (State)—New York. | Environmental policy—New York (State)—New York.
Classification: LCC GF504.N7 (ebook) | LCC GF504.N7 C36 2017 (print) | DDC 333.77—dc23
LC record available at https://lccn.loc.gov/2017016806

Cornell University Press strives to use environmentally responsible suppliers and materials to the fullest extent possible in the publishing of its books. Such materials include vegetable-based, low-VOC inks and acid-free papers that are recycled, totally chlorine-free, or partly composed of nonwood fibers. For further information, visit our website at cornellpress.cornell.edu.

For all the people and plants in New York City,
but especially Ricardo.
And in memory of Pop'Z, my favorite gardener.

Contents

Tables, Maps, and Illustrations

Tables

Maps

Figures

Preface and Acknowledgments

I work for the U.S. Department of Agriculture, Forest Service. In New York City. In the past, this has elicited bemused remarks about Smokey Bear, Central Park, and the "concrete jungle." Yet I find these kinds of comments diminishing as the awareness of and interest in urban greening grows and proliferates. Because of efforts as varied as tree planting campaigns, green roof installations, compost programs, and farmers markets, the public's relationship toward urban nature is changing. The urban forest is beginning to feel less like an oxymoron as policymakers, designers, activists, and scientists all strategize about ways to retrofit and rework our built, natural, and social environment in cities.

As a social scientist with training in human geography and urban planning who has worked at the interface of research and natural resource management praxis in New York City since 2002, I felt compelled to tell a piece of this story. The investments in parks, forests, and green space that flowed from PlaNYC, the 2007 municipal sustainability plan, felt like game changers to those of us who worked in this field. City Hall was, for a moment in time, paying real attention to trees and supporting urban forestry with large amounts of public funding. But why trees? How did other site types and issue areas connected to urban nature fare in this process? Where was the public support for the burgeoning efforts around urban farming and community gardening that I was seeing daily among activists, on the streets, and in the newspaper? This central question of which issues were embraced by PlaNYC and which were left on the cutting-room floor provided me with an entry point to examine the politics of sustainability planning and implementation.

As I conducted my doctoral research at Rutgers University, while supported by and working for the U.S. Forest Service, I saw a real opportunity to bring theoretical lenses and epistemological approaches used in human geography—including political ecology, Actor Network Theory, and post-humanist perspectives—to bear on the changing socio-nature of New York City. And New York City is far from alone in engaging in these sorts of greening and sustainability practices—global cities, mid-size cities, and shrinking cities are all wrestling with different strategies to reshape the form, function, and meaning of the urban environment. While my aim is certainly not to *predict* how these processes unfold in other cities through this single case study, I do hope that the patterns I have observed and the lessons I have learned have broader applicability to researchers, students, policymakers, and urban dwellers living and working in diverse locales. These lessons do not map one-to-one, but perhaps the detailed telling of a single case can help those working elsewhere to see where there are consistencies, and where there are important disjunctures.

The New York City Urban Field Station where I work was created in 2006 through a founding partnership between the U.S. Forest Service and the NYC Department of Parks and Recreation (DPR) and has expanded to include more than fifty collaborators. Our mission is to improve quality of life in urban areas by conducting and supporting research about social-ecological systems and natural resource management. In New York City, the Forest Service doesn't own or manage any land, and we have no regulatory authority. Our role is to conduct research and develop applications, working hand in hand with municipal, nonprofit, and private land managers. DPR manages approximately thirty thousand acres of public green space, from parks, to forests, to wetlands—as well as all of the street trees citywide. So my perspective as a researcher is fundamentally informed and shaped by this close partnership and access I have to local decision-makers and managers. As a federal research scientist, I have an obligation not only to advance theory, but also to serve the public. My role is not to offer policy prescriptions directly, but rather it is to advance knowledge that can inform sound policy and management practices conducted by others. For a geographer studying politics, policymaking, and governance, this line between engaging in critique and seeking to inform praxis without explicitly aiming to advise on policy is a delicate needle to thread. But I hope it is one that allows the reader to take in the stories and evidence presented here and reach their own conclusions about the best way forward.

This book is a work of qualitative social science that builds upon traditions of embedded, reflexive research in human geography and follows Flyvbjerg's (2001, 166) charge to "take up problems that matter to the local, national, and global communities in which we live" in order to create context-specific, practical wisdom. Having the sort of embedded access to gatekeepers, informants, and interlocutors that the Urban Field Station provided fundamentally shaped the sort of knowledge that I could produce. I see myself as neither a pure researcher nor a pure practitioner, but as one who aims to span this divide. At the same time, embeddedness is not easy. Any criticisms that I level will be met with the scrutiny, counterarguments, and rebuttal of my colleagues. In that process of dialogue, I believe that my understanding of these nuanced phenomena has only been strengthened. I hope that this book helps to continue that dialogue, pushing us further in our efforts to build sustainable, just, and resilient cities.

This book reflects the insights, guidance, input, and critique of a broad assemblage of actors. I owe much gratitude to all those who have taught me how to "see the forest for the trees" in the great complexity that is urban social-ecological systems. Erika Svendsen—since 2002 you have been a mentor, my closest research collaborator, and a dear friend—I look forward to many more years of shared inquiry, creativity, and work. Over the course of this project, Bob Lake has been a constant, with thoughtful feedback and unwavering support and advice. I am grateful for every conversation from New Brunswick to Brooklyn. All my Forest Service colleagues have been an important part of this work, with special thanks to my project leaders, current and former, Lynne Westphal, Mark Twery, and Keith Nislow. Particular thanks go to my editor, Michael McGandy of Cornell University Press, for his interest in and support of this work from when it was just a conference abstract until it was a fully realized book. Thanks also to Bethany Wasik of Cornell University Press for your attention to detail and assistance with all matters of preparing the manuscript for publication. I am also grateful to two anonymous reviewers whose feedback strengthened the rigor and clarity of the book. As always, any errors or omissions remain my responsibility.

Thank you, thank you to my current and former New York City Urban Field Station colleagues, Gillian Baine, Rich Hallett, Michelle Johnson, Dexter Locke, Heather McMillen, Renae Reynolds, Bryant Smith, and Nancy Sonti. Special thanks to Michelle for her assistance in creating the maps for this book. The NYC Urban Field Station—and this project—would not exist without our partnership with the NYC Department of Parks and Recreation. Sincere thanks to these colleagues, present and past: Novem Auyeung, Jennifer Greenfeld, Bram Gunther, Kristy King, Jacqueline Lu, Ruth Rae, Brady Simmons, Susan Stanley, and Fiona Watt—you built and continue to grow the Urban Field Station with your dedication. Thanks also to the leaders of both NYC Parks and the Northern Research Station who had the foresight and gumption to agree that New York City needed an Urban Field Station: Adrian Benepe, Liam Kavanagh, Michael Rains, Tom Schmidt, and Hao Tran. I offer gratitude to Matt Arnn who brought me into the agency as an eager college graduate, and to Jim Freund and the Princeton University ReachOut'56 fellowship for making that possible. I am grateful for my broader circle of fellow Forest Service scientists and managers focused on urban and social dimensions, particularly Marla Emery, Paul Gobster, J. Morgan Grove, Bob Haight, Sarah Hines, Beth Larry, Sarah Low, Tischa Muñoz-Erickson, Lara Roman, and Stephanie Snyder—all of whom have in some way inspired, touched, reviewed, advised on, or informed aspects of this work. Thanks to Susan Wright and the Communications and Science Delivery team for assistance with indexing.

I would have nothing to say about New York City's forests or farms were it not for the inspired folks who have been involved in greening the city over the last decades. I salute all of the hundreds of municipal agencies and workers, civic organizations and their members that are conserving, managing, monitoring, transforming, educating about, or advocating for the urban environment. There are too many to name, but special thanks to Morgan Monaco, Sue Donoghue, Andrew Newman, Edie Stone, and Lenny Librizzi. The entire Stewardship Mapping and Assessment Project (STEW-MAP) team has helped shape my thinking on urban environmental stewardship and civic engagement, particularly Dana Fisher and James Connolly. I enjoy watching our circle of STEW-MAP colleagues and collaborators grow from New York City, to Baltimore, to Chicago, to Seattle, to Philadelphia, to Los Angeles, to San Juan, to Paris, and beyond . . . STEW-MAP abides! I have learned a great deal about stewardship and urban ecology, locally and globally, from David Maddox and The Nature of Cities writers; Bill Burch of Yale University School of Forestry and Environmental Studies and Colleen Murphy-Dunning of the Urban Resources Initiative; Keith Tidball of Cornell University; Sarah Charlop-Powers and Helen Forgione of the Natural Areas Conservancy; Urban Ecology Collaborative participants near and far, present and past; and Steward Pickett, Dan Childers, and the entire Urban Sustainability Research Coordination Network.

Professors have served as key advisors and critics throughout my career. Thank you to the other members of my Rutgers doctoral committee—Rick Schroeder, Trevor Birkenholtz, Kathe Newman, and Laura Lawson. From coursework, to proposal creation, to research, analysis, and writing, you nurtured and challenged my thinking. And your comments at my defense that "this should be a book" gave me the spark to keep going. My interest in the relationship between people and their environments, and particularly the role of civil society groups in shaping urban space, was developed and honed while at MIT's Department of Urban Studies and Planning. Particular gratitude goes to my master's thesis adviser, JoAnn Carmin, and to Anne Spirn. I am also appreciative of my undergraduate thesis adviser, Julian Wolpert, who gave me my first research internship and demonstrated the breadth, depth, and possibility of a life as a geographer.

I could not produce this work without the insights and support of classmates, friends, colleagues, and family. Aimee Hess—you have been more than a friend—you have patiently helped me to refine and clarify my voice with your keen editorial eye. Isaac Gertman provided crucial design assistance—he redrew my Social Network Analysis diagrams so that they could be shared in this book. Thanks also to Carrie Grassi and Debby Scott for listening, thinking, and kvetching with me and to Amy Lerner for ongoing encouragement and feedback on this work. The Totten Fellows helped me workshop some parts of the urban agriculture case.

Thanks for being fellow explorers of urban nature—Adrina Bardekjian, Sadia Butt, Bryce DuBois, Luke Drake, Nate Gabriel, Phil Silva, and James Steenberg. Final thanks go to my family: Ricardo, mom, dad, Natalie, Dan, Alice, Hope, and Mom'Z—you are all a part of this work through your love, support, interest, feedback, and listening. All of these people and many more unnamed have helped to plant and sow this garden.

Acronyms and Abbreviations

AFN	Alternative Food Network
BBG	Brooklyn Botanic Garden
BFC	Brooklyn Food Coalition
BUG	Black Urban Growers
CDBG	Community Development Block Grant
CFH	New York City Department of Parks and Recreation's Central Forestry and Horticulture Division
CSA	Community Supported Agriculture
DCAS	New York City Department of City Administrative Services
DEP	New York City Department of Environmental Protection
DIY	Do-it-yourself
DOE	New York City Department of Education
DOHMH	New York City Department of Health and Mental Hygiene
DOT	New York City Department of Transportation
DPR	New York City Department of Parks and Recreation
DSNY	New York City Department of Sanitation
EBT	Electronic Benefits Transfer (electronic payment system for Supplemental Nutrition Assistance Program, also known as "food stamps")
FoodNYC	Manhattan Borough President Scott Stringer's *FoodNYC: A Blueprint for a Sustainable Food System*
FoodWorks	New York City Council Speaker Christine Quinn's *FoodWorks: A Vision to Improve NYC's Food System*
FPC	New York City Food Policy Coordinator
FRESH	New York City Food Retail Expansion to Support Health program
FSNNYC	Food Systems Network of New York City
GIS	Geographic Information System
GreenThumb	New York City DPR's community gardening program
HPD	New York City Department of Housing Preservation and Development
HUD	United States Department of Housing and Urban Development
ICLEI	International Council for Local Environmental Initiatives, now called Local Governments for Sustainability

LiDAR	Light Detection and Ranging—a remote sensing technique
MTTP	MillionTreesNYC Training Program
NRG	New York City DPR's Natural Resources Group
NYBG	New York Botanical Garden
NYCHA	New York City Housing Authority
NYCCGC	New York City Community Garden Coalition
NYRP	New York Restoration Project
OLTPS	Mayor's Office of Long-Term Planning and Sustainability, now called Mayor's Office of Sustainability
OMB	New York City Office of Management and Budget
PlaNYC	PlaNYC2030: New York City's Long Term Sustainability Plan from 2007
PlaNYC 2.0	2011 update to PlaNYC
PROW	Public Right of Way (streets and sidewalks)
SAB	Sustainability Advisory Board (to PlaNYC)
SNA	Social Network Analysis
STEW-MAP	Stewardship Mapping and Assessment Project (U.S. Forest Service project)
STRATUM	Street Tree Resource Analysis Tool for Urban Forest Managers (U.S. Forest Service model)
TNY	Trees New York
TPH	New York City DPR's Trees for Public Health program
TPL	Trust for Public Land
UFORE	Urban Forest Effects model (U.S. Forest Service model)
USDA	United States Department of Agriculture
UTC	Urban Tree Canopy (U.S. Forest Service model)

CITY OF FORESTS,
CITY OF FARMS

Introduction

JUXTAPOSING URBAN FORESTRY AND AGRICULTURE IN THE PLANYC ERA

On November 20, 2015, municipal officials and nonprofit staff planted the 1,017,634th tree of the MillionTreesNYC campaign at Joyce Kilmer Park in the Bronx. Many policymakers consider the campaign, which began in 2007, one of the crowning achievements of New York City Mayor Michael Bloomberg's long-term sustainability plan, PlaNYC 2030. Inspired by the Department of Parks and Recreation (DPR)'s quantified arguments about the benefits of trees, PlaNYC committed approximately $400 million in capital funding to expand New York City's urban forest—on streets, in landscaped parks, and in wooded areas. This was the largest such investment in urban parkland since the Robert Moses era of the mid-twentieth century. At the same time—and in response to the interests of its founder, entertainer Bette Midler—the local nonprofit New York Restoration Project (NYRP) sought to plant one million new trees citywide. These two organizations joined together as MillionTreesNYC, a public-private partnership that would plant across all land jurisdictions, leverage private dollars, and engage the public as supporters and stewards. With celebrity-backed fundraisers, business-like management of the Bloomberg era, and massive volunteer planting days, this campaign has inspired urban tree planting efforts across the globe, from New Jersey, to Salt Lake City, to London.[1] These campaigns are part of a larger movement toward investments in green infrastructure to enhance urban environments and quality of life.

In contrast, urban greening practices such as urban agriculture, community gardens, and other interventions into local food systems received *no mention* in the 2007 version of PlaNYC. Yet gardening and farming had been vibrant areas of civic engagement and grassroots activism for decades. Since the 1970s, New York City has been home to one of the largest networks of urban community gardeners in the world, involving thousands of individuals who manage hundreds of gardens citywide. In the mid-2000s, a new wave of interest in seasonal and organic food, local farms, farm-to-table restaurants, community-supported agriculture (CSA), and farmers markets as part of Alternative Food Networks (AFNs) has flourished. At the same time, the twin crises of hunger and obesity have increased urgency around the issues of food security, healthy food access, and food justice. From 2009 to 2011, hundreds of civic actors worked with municipal officials to craft local food visions and plans that articulated both challenges and potential solutions. Food issues were subsequently incorporated into the 2011 update to PlaNYC, but somewhat marginally. Despite a handful of new laws and programs, New York City's urban agriculture policies are not as far-reaching as the comprehensive food strategies of places like Vancouver or Toronto and nowhere near the resources and attention devoted to urban forestry via MillionTreesNYC.[2]

Why were urban forestry and agriculture treated so differently in the sustainability plan? And, more broadly, how is nature constructed in the city via politics, policymaking, and governance? Investing in urban green spaces—including parks, trees, waterfronts, and brownfields—has become a crucial component of many municipal sustainability efforts. Understanding *how and why* the sustainability agenda is set provides crucial lessons to scholars, policymakers, and activists alike as they engage in the greening of cities. In the urban forestry case, we see mayoral commitment operationalized through a savvy municipal agency (DPR), coupled with an influential nonprofit (NYRP). In the urban agriculture case, we see an initial lack of mayoral interest, but food and agriculture were slowly brought into the municipal fold via the work of other public officials and civic activists. The stakeholder networks vary dramatically across these two issue areas. Urban forestry is centralized around the key, professionalized public agency of DPR, whereas urban agriculture is a more polycentric, civic-led network with a long history as a social movement. Through these distinct cases, we can learn whose voice is heard and seen as legitimate in setting New York's urban greening agenda. And we can inquire what gets lost when certain voices or issues are bracketed out of a plan that aimed to support the green growth of the city. By examining how different interests and power dynamics play out in setting the urban environmental agenda in New York City, we can reflect on those same dynamics in other cities and towns. Through this reflection, planners can work to develop strategies and tactics for steering their plans toward more inclusive and just processes and ends.

At the same time, the construction of urban nature is driven not only by political actors and networks but is also influenced by storylines and nonhuman actors. First, the differences in PlaNYC's narratives about urban forestry and agriculture reveal that decision-makers often privileged a commodified view of nature. PlaNYC portrayed trees as a sound investment because they provide ecosystem services at a favorable cost/benefit ratio.[3] In contrast, many of the critiques of urban agriculture from policymakers related to the lack of quantified metrics about the benefits of urban farming and community gardens. Activists and academics have worked to address that gap through citizen science and research. Second, physical constraints and abilities of nonhuman actors—trees, sidewalks, buildings, lots, and farms—also shape natural resource management practices.[4] In New York City, any intervention into urban nature occurs in the context of the highly developed city, with its condition of lack of space. One million new trees could be inserted into the already-existing built environment without competing with other residential, commercial, or industrial uses. In contrast, urban agriculture generally requires sites that are suitably wide, sunny, and flat; such sites are becoming increasingly scarce in the growing city, which has led growers to experimentation with alternative sites and techniques, including rooftop farms and temporary plantings. Limitations of space, cost, and productivity have pushed urban agriculturalists into thinking at broader scales and forming regional alliances in support of the entire food system, from production and processing to consumption and post-consumption.

This book explores the politics of urban greening through an investigation of the creation and implementation of PlaNYC2030 in New York City from 2007 to 2011 and the implementation of MillionTreesNYC from 2007 to 2015. PlaNYC was created as a plan for the "physical city," with particular emphasis on municipal infrastructure, in light of projections that New York City's population would reach nine million by 2030 (City of New York 2007). While other issue areas were included in the sustainability plan—ranging from housing, to transportation, to water quality, to energy, to air quality—I focus on transformations of public (and private) green space as part of a broader turn toward urban greening. Drawing upon in-depth interviews, discourse analysis, social network analysis, and participant observation, I compare the cases of urban forestry and urban agriculture to reveal how politics, discourse, and natural resource management practices both drive and respond to institutionalization within the sustainability plan.[5] I explore key decision-makers, constituencies, turning points, and policy levers that explain the scope of the plan. Moreover, because the plan was issued in 2007 and updated in 2011, I describe how policymaking is malleable or incremental—even while being comprehensive and long term. Thus, I show that agenda-setting is not a single moment in time but rather an ongoing negotiation among intertwined actors (both human and nonhuman). And I describe how the seemingly innocuous stage of implementation remains highly political, with important consequences for how urban nature is produced. In this stage, bureaucrats and advocates have room to maneuver in pursuit of their aims, including those that were not formally stated in the plan—such as social justice.[6] I also identify a range of political and quotidian practices that shape the creation and experience of urban nature beyond the bounds of the plan. Understanding how urban nature is constructed in a global city can provide insights to sustainability planning and natural resource management that are occurring in hundreds of other cities across our urbanizing world.

In light of this case, planners, policymakers, and activists can consider the ways in which the sustainability mantle gets taken up, altered, and translated through the processes of goal setting, writing, and implementation in other cities across the world. Urban greening creates material transformations of the city but also is dictated by political processes and discourses that shape its parameters. With this understanding in mind, planners can find different kinds of apertures and spaces for engagement around the sustainability agenda. For example, we can broaden the physical sites that we consider as our policy objects, including rooftops, underused street intersections, and parking lots. We can take new approaches to policy discussions by working with diverse platforms for communication, such as participatory budgeting, crowd-sourcing ideas, and social media. We can constantly reflect on who is missing from the dialogue and what further can be done to hear and address their concerns. Finally, we can interrogate our own assumptions of what sustainability *is,* who it is for, and what it looks like in the urban terrain—so that we do not exclude "future ideas" as not being "actionable," but instead identify novel pathways and innovations that can lead us toward those future states.

Local Sustainability Planning on the Rise in New York and Beyond

Whether due to the pressures of urbanization, growth, and climate change, a "mainstreaming of environmental values" (Keil and Boudreau 2006, 49), or trends in policymaking amongst competitive cities, local sustainability planning efforts are on the rise (Harvey 1989; Sadik 1999; Gibbs and Jonas 2000; Prytherch 2002; McGranahan et al. 2005; Jonas and While 2007; Pierce and Johnson 2008; Finn and McCormick 2011). While historically sustainability efforts were global in reach and were often advanced through the United Nations and its related meetings, other more recent efforts are independent local policy initiatives, albeit ones that are connected through networks of municipal leaders.[7,8] A 2007 *Time* magazine cover story about this rescaling featured Michael Bloomberg, then mayor of New York City, and Arnold Schwarzenegger, then governor of California, with the headline "Who Needs Washington?" In addition to comprehensive sustainability plans, city leaders are developing large-scale investments in tree canopy and green infrastructure as well as local food policies (NYC Council 2010; Pincetl 2010).

While urban sustainability plans are burgeoning amongst policy practitioners, many scholars critique these efforts on scalar and structural grounds. They argue that the notion of a "sustainable city" is a farce—as it does not adequately consider the hinterlands and global commodity chains on which city life depends. Simply drawing a line around a city and focusing on the processes within that boundary does not remove the impacts of urban lifestyles that are borne outside the line (see, e.g., Swyngedouw 1996; Swyngedouw and Heynen 2003; Braun 2005). Geographers criticize local sustainability planning as supporting capitalism, as a form of greenwashing, or as a "flanking mechanism" to neoliberalism—arguing that current sustainability plans do not challenge the fundamental political-economic structures currently at work (Gibbs and Jonas 2000; Brenner and Theodore 2002; Jonas and Gibbs 2003).

Acknowledging these legitimate critiques, we can nonetheless view urban sustainability plans as opportunities for incremental, progressive change. By examining practices that local decision-makers deem related to sustainability, we can investigate how these practices came about, why they took the form they did, and how they shape the urban sphere. Overall, this book adopts a "politics of hope" toward the possibility of change in the urban sphere through greening (Amin and Thrift 2002, 4). Embracing the politics of hope means seeing space to maneuver and to create new futures and possibilities, in spite of structured inequalities, institutional rigidity, and capitalist social relations. It means being open to the unexpected encounter, to unpredictability, and to change. I argue that as the physical form of cities changes, we can create public spaces that encourage sociability, new interactions, and creative improvisation among residents that, in turn, have the potential to influence further changes in the future of the city. The plans and programs analyzed herein do not propose radical reworking of our political economy, but they do create shifts in the physical form of the city, as well as shifts in the institutional arrangements that support those changes.

PlaNYC2030 was New York City's long-term sustainability plan that was created in 2007 under the Michael Bloomberg mayoral administration. Bloomberg created an Office of Long-Term Planning and Sustainability (OLTPS) to oversee PlaNYC. Long-term sustainability planning was institutionalized via a local law requiring the plan to be updated every four years, along with reporting on an annual basis. PlaNYC set goals and targets to improve local quality of life in terms of land, water, transportation, energy, air, and climate change (City of New York 2007). It largely focused on goals that were achievable through municipal action alone, by taking on issues that were within the jurisdiction and mandate of city agencies.[9] PlaNYC committed substantial city government capital and human resources to these aims. As such, it is an example of "roll out neoliberalism" that involved not outright state retreat, but rather new forms of governance institutions and new investments (see Peck and Tickell 2002). Notably, it was created in a time of municipal surplus and economic growth, just prior to the global financial crisis of 2008. The update to the plan, released in 2011 and informally dubbed "PlaNYC 2.0," reflected the changed economic context through more modest goals and lack of commitment of new financial resources.

The main aim of PlaNYC was to accommodate population growth in a way that ensured the livability and economic competitiveness of New York City. Indeed, the plan initially evolved out of New York City's bid to host the 2012 Olympics (ICLEI 2010b; Fuchs 2011). It started from demographic projections that showed the city increasing by one million residents by 2030 and proposed infrastructure upgrades, investments, and policies to manage that change (City of New York 2007). While its scope and complexity expanded from its initial focus on land use and infrastructure, PlaNYC was not fundamentally focused on social services, education, or employment programs. This sort of service delivery was, for the most part, bracketed out of this sustainability plan. Moreover, nothing about PlaNYC was progressive to the point that it would alienate business interests; and critics claim that it served the economic elite. The composition of the plan's advisory board is illustrative: elected officials (two members), business/real estate interests (six), environment, community advocacy, and planning (six); academia and philanthropy (two); and labor (one) (ICLEI 2010b, 20–21).

At the same time, the policies represented real advances in supporting local environmental quality. Goals included improvements in water quality and increases in urban tree canopy, number of parks, and affordable housing. The plan has been widely praised in pragmatic planning, management, and local government circles. At a public panel on New York as a Sustainable City, geographer William Solecki (2011) called it a "transformative document" and said that New York City is being held as a "beacon of urban sustainability" nationally and globally. There was a fairly broad level of support and enthusiasm for the plan among local environmental advocates during its implementation. Often where there was critique, it was over a desire to expand the plan into new arenas (such as urban agriculture and solid waste management) rather than to condemn it wholesale. This book attends to the plan's evolution in the face of critique, implementation, and subsequent revision.

The plan was intended to drive action rather than solely advise or recommend. PlaNYC was *not* a formally approved city plan; it was not demanded by the electorate, nor was it the project of the city council. It was a set of strategic initiatives crafted by City Hall to be part of Bloomberg's environmental legacy, funded primarily with capital expenditures, and implemented through city executive agencies. Bloomberg's leadership has been critically analyzed as a quintessential example of neoliberal governance, with his aims to "run the city like a business" and treat citizens as customers (Brash 2011). PlaNYC's initial process was publicly criticized by some for being too top-down and not participatory enough in its formulation, with public meetings functioning as no more than tokenism (see, e.g., Barrett 2007; Mandelbaum 2007; Angotti 2010; see also Rosan 2011). Others countered that the construction of the advisory board and the public outreach process surrounding the plan made it "the most inclusive process in New York City in many years" (Fuchs 2011; see also ICLEI 2010b). This book occupies a space between these claims of criticism and celebration. It does not ignore the important critiques of PlaNYC's process or scope, but it also seeks to understand the effort as a form of "actually existing sustainability" (Krueger and Agyeman 2005). It illuminates *how* Bloomberg, City Hall, and OLTPS's bureaucratic staff worked with a wider network of partners, including other municipal agencies, local civil society groups, and even nonhuman actors to craft and implement PlaNYC's urban forestry and urban agriculture agendas, to which I now turn.

Urban Forestry

MillionTreesNYC was one of 127 initiatives in PlaNYC and has generally been celebrated as one of the successes of the municipal plan (see, e.g., ICLEI 2010b). The campaign aimed to plant and care for one million new trees in New York City by 2017.[10] The campaign advanced the PlaNYC goal of "reimagining the public realm" (i.e., the sidewalk and other public spaces) with a target of planting in every available street tree planting site and an initiative to reforest two thousand acres of "natural area" parkland citywide (City of New York 2007).[11] DPR is a highly advanced urban natural resource management agency, and it was well positioned to advocate for these goals in PlaNYC. But the campaign's origins are located in both government and civil society; simultaneously and separately from the plan, Bette Midler, NYRP's founder, became interested in the idea of planting a million trees in New York City. NYRP is a greening nonprofit founded in 1995.[12] The resulting campaign emerged via a carefully negotiated, public-private partnership between DPR and NYRP. Once the goal of planting one million trees was jointly embraced and embedded in the plan, city government relied on NYRP for its capacity to support a higher-profile campaign that would last beyond Mayor Bloomberg's tenure, by both fundraising and planting trees on private property. Conversely, NYRP relied upon DPR for its expertise in the large-scale operation of urban forestry—from tree procurement, to planting, to maintenance. In addition, leaders of the campaign sought to build a broader local movement around urban forestry, and many other stakeholders were crucial. Key private funders (including Toyota, BNP Paribas, Home Depot, Bloomberg Philanthropies, and David Rockefeller) supported the campaign financially, and an advisory committee of 400 individuals from 109 organizations contributed expertise in urban forestry and stewardship.

The campaign asserted that one million additional trees were good for the livability of the growing city, building upon decades of research from the U.S. Forest Service, universities, and the professional forestry and arboriculture sector about the benefits of the urban forest (see, e.g., Anderson and Cordell 1988; McPherson and Simpson 2002; Wolf 2005; Donovan and Butry 2010; Nowak et al. 2010). A report analyzing the costs and benefits of street trees—especially their potential to increase real estate values—helped convince the mayor and his staff to include tree planting goals in PlaNYC (Peper et al. 2007). Later, during implementation of the campaign, a large-scale print ad featured images of trees and people in the city, touting trees as "Zen masters" and "exercise partners." While municipal leaders were convinced by economic arguments, the campaign messaging included a broader scope of quality of life, ecosystem health, and human health concerns. The multiple benefits of trees provided flexibility, such that a range of arguments could be marshalled to support investment in the urban forest.

The campaign created innovative volunteer programming but did not follow a community forestry ethos whereby citizens have control over natural resource management decisions. DPR and NYRP offered environmental education programs with a particular focus on schoolchildren, seeking to increase ecological literacy and cultivate the next generation of urban environmental advocates. Stewardship opportunities were created for thousands of participants on city streets, in parks, in "natural areas" (or forests), at school grounds, on other institutional grounds, and in private yards via tree giveaway programs. At the same time, the funding for MillionTreesNYC—which focused primarily on planting and not on maintenance—meant that the campaign had a real need to engage volunteer labor in planting trees in parklands and in the stewardship of street trees.

The infusion of PlaNYC capital funds, the establishment of a numeric planting goal, and the institutionalization of the campaign all dramatically altered the practice of urban forestry in New York City. DPR changed its procedures for tree procurement, planting, and contract management and drove changes to zoning rules related to street trees and trees in parking lots. DPR also developed a block planting strategy that concentrated first in "Trees for Public Health" neighborhoods—areas that were low in street tree stocking levels and had high rates of childhood respiratory incidence—demonstrating a commitment to environmental justice. Similarly, NYRP scaled up to have a citywide urban forestry campaign, whereas previously it had focused on parks and gardens in a few targeted neighborhoods. NYRP forged relationships with corporate donors and volunteers, planted on sites such as New York City Housing Authority (NYCHA) public housing grounds, and engaged in large-scale marketing and tree giveaway campaigns. Finally, MillionTreesNYC changed throughout its implementation, particularly in response to the 2008 global economic recession, leadership turnover at both DPR and NYRP, and the maturation and learning of the campaign itself. Although the campaign formally ended in 2015, many of the procedural and programmatic changes brought about by DPR will continue to influence the practice of urban forestry in New York in perpetuity going forward.

Urban Agriculture

In contrast, urban agriculture was entirely absent from the first edition of PlaNYC. There was no explicit goal related to local food production, and neither urban farms nor community gardens were mentioned in the plan, despite a long tradition of gardening in New York City. DPR's GreenThumb program is one of the largest community garden programs in the country, with more than five hundred gardens and supporting approximately twenty thousand gardeners citywide with access to land, technical assistance, and materials, such as soil, compost, and plants (GreenThumb 2010). Many of these gardens were created by residents in the 1970s and 1980s in the time of the city's last financial crisis (Lawson 2005). Similarly, the first farmers market in New York City dates to 1976 (GrowNYC 2010). Since the 1960s, hundreds of community-managed gardens have also been located on NYCHA public housing grounds. School gardens on public and private lands have existed in New York City for over a century (Lawson 2005), but a new public-private partnership of the nonprofit GrowNYC and the Department of Education has encouraged a boom in school gardening.

In addition to these wide and deep roots, the 2000s brought a renewed wave of interest in urban agriculture and local food systems. New York City now has commercial rooftop farms, urban farms, greenhouses, backyard chickens, beekeeping, and heightened media attention toward these sites (Salkin 2008; Ryzik 2009; Cardwell 2010; Stein 2010; Wells 2010; Cohen et al. 2012). CSAs are on the rise locally and nationally, with more than 137 in the Just Food network for New York City as of 2016 (Ostrom 2007; Just Food 2016). Similarly, there are now more than fifty Greenmarket farmers markets citywide (GrowNYC 2010). There is also an effort to create a larger-scale, permanent farmers market akin to the Ferry Building Marketplace in San Francisco, as well as a wholesale farmers market to serve the restaurant industry (Severson 2008). This coincides with rising national attention to food provenance and health issues through books like

Michael Pollan's (2006) *Omnivore's Dilemma;* documentary films like *Food Inc., Fresh,* and *Supersize Me;* and the work of celebrity chefs including Berkeley's Alice Waters of Chez Panisse and New York's Dan Barber of Blue Hill (Salkin 2008). This has coalesced with a contemporary interest in do-it-yourself (DIY), crafts, and homesteading that has led to a dynamic local food movement in New York City that includes not only gardens and farms, but also home brewing, cheese-making, kombucha making, and so on. At the same time, a diverse movement with more explicitly political aims around food justice and food sovereignty has also been maturing over the course of the last few decades, not only in New York City but nationally and globally. This movement focuses on strengthening community self-determination around food, with an emphasis on access to fresh, healthy, affordable, and culturally appropriate food for all (see also Reynolds and Cohen 2016).

Yet food was seen as outside the scope of PlaNYC at the time of its writing in 2007. While the Bloomberg administration did create policies and programs around healthy food *consumption,* it initially failed to engage in broader conversations around food *production* and the regional "foodshed." Policymakers critiqued the lack of data to demonstrate urban agriculture's productivity and raised concerns over the lack of sufficient space in the developed city to achieve any significant levels of production. Some policymakers used the concept of the rooftop farm symbolically as a win/win for adaptive reuse and green urbanism that did not compete with other ratable uses. But for activists focused on food security, justice, and sovereignty, the rooftop greenhouse selling its greens to Whole Foods' wealthy clientele symbolized everything wrong with urban agriculture. There were divisions between a new wave of young, often white, "foodies" and farmers and the prior generation of community gardeners who were often low-income people of color. At the same time, these tensions provided productive fodder for discussion, critical self-reflection, and community organizing to help build an inclusive food movement. Activists and allies sought to build large coalitions that could span the issues of healthy food access, obesity, diabetes, hunger, sustainable urbanism, and regional farm conservation.

Organizing and advocacy led to the development of a series of reports and plans that laid out visions for transforming local food systems. Individuals, community-based groups, and nonprofits worked in coalition with two then-mayoral candidates, former Manhattan borough president Scott Stringer and former New York City Council Speaker Christine Quinn, to bring food and agriculture into the municipal policy arena via these plans.[13] In response to public critique of the first edition of PlaNYC, as well as the changed political landscape around food issues, urban agriculture subsequently was included in the 2011 updated PlaNYC 2.0. The plan had a brief crosscutting section on food and a specific initiative on enhancing urban agriculture and community gardening citywide (City of New York 2011a). It made commitments to conduct a citywide inventory of potential sites for food production; to create 129 new gardens on NYCHA grounds; to develop 150 school gardens through Grow to Learn; to increase membership in GreenThumb community gardens by 25 percent; and to create five new farmers markets. Yet because PlaNYC 2.0 was released in 2011 under very different economic conditions, it made *no new municipal budget commitments toward these goals*. Tracing the rollout of the Stringer, Quinn, and PlaNYC documents over time reveals the way in which narrative concepts, ideas, and specific proposals are reiterated and legitimized as food and agriculture become embedded—however nominally—in municipal policymaking arenas. Thus we begin to see the possibility for incremental, local change in urban agriculture and food policies. It is important to note that—given the glocal and transnational nature of the food movements—these municipal visions and plans did not occur in isolation. They also were influenced by and intertwined with organizing at the national and global levels—but these scales are largely beyond the purview of this book.

Research Questions and Perspectives on the Construction of Urban Nature

Using New York City in the Bloomberg era as a case study of sustainability policymaking and implementation, this book investigates three main sets of questions about the *governance and politics, discourse*, and *materiality* of urban nature. These three different lenses are crucial for understanding sustainability planning as a process that unfolds through the strategic interplay of actors, the deployment of different narrative frames, and the mobilizing and manipulation of the physical environment (including other living, nonhuman entities). It is important to remember that these concepts are mutually constitutive and overflow the categorical tidiness with which they are presented. Nonetheless, the next three subsections explore these themes sequentially. Multiple theoretical approaches from human geography, urban studies, political science, and sociology are useful as entry points for answering these questions and for exploring the complex and contested construction of urban nature.[14] Overall, social theory encourages us to consider who governs urban nature, how, why, and to what effect.

Governance and Politics

The first set of research questions guiding this book asks who are the actors involved in the politics and governance of urban forestry and urban agriculture in New York City; and how do they participate in these practices? In particular, what are the roles of the mayor and bureaucratic officials in policymaking and of civil society in a government-led sustainability plan? At the same time, which actors are bracketed out of these processes? How do civil society groups foster urban forestry and urban agriculture practices outside of the context or confines of a municipal plan? How do these actors work together through networked and hierarchical structures to make on-the-ground change in the policies, programs, and form of the city?

In order to address these questions of how environmental governance occurs, I examine the multiscalar interactions among the state, civil society, and private sector as they attempt "to alleviate recognized environmental dilemmas" (Davidson and Frickel 2004, 471). Some clarifying definitions are important here. Jordan (2008, 21 quoting Kooiman 1993, 2) defines *governing* as "those social activities which make a 'purposeful effort to guide, steer, control, or manage (sectors or facets of) societies,'" while *governance* is "the patterns that emerge from the governing activities of social, political, and administrative actors." If there is one overarching trend in the governance literature, it is this attention to nonstate actors involved in the practices of governing—which we clearly see in the New York City cases explored herein (see also Swyngedouw 2005). Debates continue as to whether or not this expansion of actors is a good thing.[15]

In looking at the expanded set of actors involved in governance, we can examine state-led or top-down efforts (Skocpol 1985); civil society–led, or bottom-up efforts (Piven and Cloward 1979); as well as networks that "can blur, even dissolve, the distinction between state and society" (Rhodes 1996, 666). Networks are a governing structure for allocating resources and coordinating decision-making that can be contrasted with other structures, such as hierarchy or markets (Rhodes 1996; Jordan 2008). Rocheleau and Roth (2007) argue that networks serve as metaphors, models, and theories to a broad set of social and natural sciences.[16] We think differently about how governance happens if we envision it more as a knotty, polycentric set of interactions than as an orderly, top-down pyramid or an inverted, bottom-up pyramid. This book embraces a network approach. According to Kjaer (2009), however, we should not think of networks as functioning in the absence of hierarchy. Indeed, participants in governance networks have different resource bases, degrees of power, and interests, all of which can lead to conflict. It is therefore important to employ the network-governance concept with a critical eye—and to supplement it with perspectives that more squarely take on the question of power.

Urban politics research has evolved in the United States through a decades-long discussion of *who has the power to govern the city*.[17] Urban regime theory has made a strong contribution to understanding the ways in which reasonably stable partnerships, political alliances, and coalitions between the state and—in particular—the private sector shape policymaking at the local scale (Elkin 1987; Stone 1989, 1993; Mossberger and Stoker 2001; Mossberger 2009). One of the lasting conceptual contributions is Stone's notion of a social production mode of power, which is defined as "'power to' accomplish goals rather than 'power over' others" (Stone 1989, 229, quoted in Mossberger 2009, 43). Stone says, "People who know one another, who have worked together in the past, who have shared in the achievement of a task, and who perhaps have experienced the same crisis are especially likely to develop tacit understandings. If they interact on a continuing basis, they can learn to trust one another and to expect dependability from one another" (Stone 1989, 4). This form of power focuses on trust, reciprocity, and coalition-maintenance and is a useful concept to bring to the examination of networked governance in urban environmental politics.

Throughout Stone's account, the mayor remains the central figure involved in creating lasting coalitions. Numerous scholars have pointed out that urban regime approaches do not give sufficient attention to the role of civil society or public bureaucrats—and have brought these actors into political accounts (Ferman 1996; Pincetl 2003; Martin 2004; Kjaer 2009). The technical formulation and implementation of policies necessitates the participation of numerous, unelected bureaucrats throughout the agencies of the state—and it is often these bureaucrats that serve as the point of interface with other stakeholders (Brecher et al. 1993; Keil and Boudreau 2006). Finally, while urban regime theory notes that mayors and business elites have privileged places in governing coalitions, it does not turn its attention directly to the way in which capitalism itself shapes urban governance (Lauria 1997; Painter 1997) or creates the presence of a "growth machine."[18]

The growth machine thesis sought to articulate a middle way between explanations of urban politics that missed structural factors and the relatively actorless accounts of urban geography favored by traditional Marxists (Molotch 1976; Logan and Molotch 1987; Logan et al. 1997). Logan and Molotch (1987) described the production of urban inequality (within and between cities) as a competition of use and exchange values that is led by a particular set of landed elite actors and institutions that can benefit from growth. Thus, they argue that the question of *whether or not to grow* is not on the agenda of urban politics. Instead, local politics determines *how to grow*, with land-use planning and development dominating local politics (Logan et al. 1997). They offer a detailed analysis of political elites, the rentier class, the media, utilities, and educational and cultural institutions, all of which benefit from the "city as growth machine" (Logan and Molotch 1987). Overall, twenty years of research on growth machines found that "the principal effect of growth machines is to bend the policy priorities of localities toward developmental, rather than redistributional goals," although this may be starting to change with the more recent era of smart growth and sustainability planning (Logan et al. 1997, 605; Jonas and Gibbs 2003).

Scholars of urban politics wrestle with the questions of who governs, via what institutional mechanisms, and in what contexts? Despite their different lenses, scholarship from urban regime and growth machine approaches both offer careful analysis of members of governing coalitions—often placing particular emphasis on the mayor, local business interests, and the electorate. This book builds upon urban regime theory, while seeking to address its relative blind spot toward the role of civil society by placing attention on the role of nonprofits and community groups in local environmental policymaking and natural resource management. The role of bureaucrats is also highlighted because of their participation in complex, technical, and normative acts of policymaking. In response to the growth machine thesis, we can investigate whether and how sustainability plans seek to manage, accommodate, or alter growth.

At the same time, this book brings insights from governance theory by seeing governing coalitions less as a fixed set of strategic players and more as a fluid form of *networked governance*. But I note that this network always exists alongside other governing structures, including hierarchy. Use of the network metaphor must not ignore issues of power, inequality, exclusion, and accountability. I see compatibility between the concept of networked governance and the social production mode of power—whereby coalitions are actively built and maintained and are valuable once established. Both encourage an analysis of policymaking as a range of active practices that unfold over time, which is precisely the approach that I take to the analysis of PlaNYC. I offer an extended view of policymaking, which includes goal setting, implementation, and revision—interrogating the causes and implications of the changes in scope and emphasis between the first (PlaNYC, issued in 2007) and the second (PlaNYC 2.0, issued in 2011) edition of the plan.

Discourse

Politics operates through domains other than just the visible interplay of strategic actors; it is also discursively constructed, materially tethered, and historically and geographically contingent. So the stories we tell, the physical things that comprise the city, and the specific histories of a city all make up that *place*.[19] Municipal agencies creating long-term plans not only shape policy trajectories but also reinforce particular ideologies about the city, the environment, growth, and capitalism. Thus, the second set of research questions guiding this book asks what discourses of the environment and society are created and deployed in sustainability plans (PlaNYC and PlaNYC 2.0), what are the ideological assumptions behind these claims, and who stands to benefit or lose from their advancement.[20] I pay particular attention to how ideologies of environmentalism, sustainability, and neoliberalism overlap or compete. And I explore whether narratives of local sustainability are utilized in advancing pro-growth agendas and interurban competition. Through this analysis we can see what version of sustainability is being advanced.

Discourses about nature and the city are actively in use in the realm of planning and politics (see, e.g., Harvey 1996; Fischer and Hajer 1999; Lake 2003; Kaika 2005; Keil and Boudreau 2006). Myriad claims have been made by planners, resource managers, and scientists about the properties of the urban forest, often focusing on the benefits of improved air quality, energy savings, enhanced neighborhood walkability, increased real estate value, and mitigation of urban heat island effect (see, e.g., Nowak et al. 2010 for a review of research on urban forests). In terms of urban agriculture and community gardens, claims range just as broadly—with perhaps a greater emphasis on the social dimensions, including access to fresh produce, redressing of food injustice, provision of neighborhood open space, educational opportunities, and the importance of community-based resource management, to name just a few (see, e.g., Schmezlekopf 1995; Ferris et al. 2001; Twiss et al. 2003; Tranel and Handlin 2006; Wakefield et al. 2007; Brdanovic 2009; Murphy-Dunning 2009; Stone 2009; Svendsen 2009).

In order to analyze such claims, we must examine the ideological assumptions, narrative frames, and communication strategies—including the strong emphasis on quantification and commodification—in urban planning and natural resource management spheres. Davidson and Frickel (2004, 477) note that social scientists should "turn their attention to the processes of environmental claims making—how social and political understandings of nature and environmental problems are crafted, contested, and legitimated." Going further, Harvey (1996, 182) asserts, "All ecological projects (and arguments) are simultaneously political-economic projects (and arguments) and vice versa. Ecological arguments are never socially neutral any more than socio-political arguments are ecologically neutral." Thus, environmental politics *is* discursive, a struggle amongst competing framings and definitions of the environment (see also Hajer 1995).

One almost cannot discuss contemporary urban or environmental life without encountering the concept of *neoliberalism*. Neoliberalism, an ideology that emphasizes markets and individual responsibility as the means for achieving efficient outcomes in a wide variety of domains, has risen to dominance particularly since the 1970s (Gibbs and Jonas 2000; Brenner and Theodore 2002; Brand 2007). The relationship between neoliberal ideology and urban areas has been thoroughly explored by scholars. Brenner and Theodore (2002, ix) claim, "The point is not only that neoliberalism affects cities, but also that cities have become key institutional arenas in and through which neoliberalism is itself evolving." Peck and Tickell (2002) argue that neoliberalism and entrepreneurialism have co-evolved, with neoliberalism naturalizing growth, market logics, and competition. They offer the important clarification that neoliberalism does not always involve the "roll back" or retreat of the state as per Jessop (2002) but can also include the "roll out" of new forms of governance institutions (Peck and Tickell 2002).

Neoliberal ideology also shapes views of nature and the environment—however, there is debate in the literature about what implications this has for urban sustainability efforts. Scholars and activists alike critique neoliberalism as taking an instrumental or commodified view of nature as primarily in service to markets. Given this, McCarthy and Prudham (2004) note a paradox that modern environmentalism co-evolved alongside neoliberalism in the post-industrial era, but at the same time, environmentalism can stand as a political challenge to neoliberalism. Perkins (2009) argues that local environmental groups support the dominant or hegemonic neoliberal mode.[21] Conversely, Bulkeley (2005, 889) claims that urban sustainability can sometimes be an "alternative to . . . the neoliberal economic project." Reflecting a middle way, Keil and Boudreau (2006, 41) use a case study to demonstrate that the neoliberal regime in Toronto during the 1990s had the unexpected side-effect of producing a strengthened urban ecological agenda, which led to the development of Toronto's environmental plan.

In sum, assumptions we have about nature—and the role of the state in managing it—fundamentally shape our urban sustainability policies and natural resource management practices. And the discursive realm is connected to material transformations of place: "The material production of environments is necessarily impregnated with the mobilization of particular discourses and understandings (if not ideologies) of and about nature and the environment" (Heynen et al. 2006, 7; see also Mele 2000). In this book, I analyze the claims related to urban forestry and urban agriculture to reveal the common logics, patterns, and tropes at work. Commensurate with the normative aims of political ecology, this analysis seeks to reveal the often unstated ideological assumptions upon which local practices rest and to interrogate *who might benefit from these claims*. Indeed, Heynen et al. (2006, 9) argue that from an urban political ecology perspective, ". . . there is no such thing as an unsustainable city in general, but rather there are a series of urban and environmental processes that negatively affect some social groups while benefitting others (See Swyngedouw and Kaika 2000). A just urban socio-environmental perspective, therefore, always needs to consider the question of who gains and who pays. . . ." By examining discourses that are put forth in environmental policies and that make recommendations about the socio-natural environment, this analysis seeks to explore the inherent linkages among the discursive, the political, the institutional, and the material.

Materiality

The final set of research questions attends to the material qualities of the natural environment to ask: How do the physical constraints and abilities of biotic and abiotic "actants," including trees, plants, sidewalks, buildings, roads, and vacant lots, shape sustainability plans and resource management practices? What are the impacts of historic changes in the form of the socio-natural environment on current policies and practices? These questions encourage us to consider the role of nonhumans as active agents, not just passive substrates, in shaping our urban environments. Scholarship with a material turn as well as approaches like Actor Network Theory (ANT) and assemblage geography are grappling with new ontologies, epistemologies, methodologies, and ways of writing that change how we make accounts of phenomena. Via these approaches, cities are seen as hybrid matrices of "urban socio-nature" that are always politically contested entanglements of humans and nonhumans.

Heynen and others (2006, 2) claim that one of the conceptual contributions of urban political ecology is to "re-nature urban theory" by contesting the division between nature and society and examining urbanization as a driver of environmental change. Political ecology approaches have been applied to a number of different questions and cases in urban settings of the global north, including suburban lawns (Robbins and Sharp 2003; Robbins 2007), urban forestry (Heynen 2003), water politics and water systems (Swyngedouw 2004; Kaika 2005), urban parks (Brownlow 2006), and New York City's socio-natural history (Gandy 2002). These analyses demonstrate that nature has been and is constantly being transformed by humans (Swyngedouw and Heynen 2003) and that "there is nothing *unnatural* about New York City" (Harvey 1996, 186 emphasis original).

Recent approaches from assemblage geographies and material geography further help to contest the nature-society binary via new analytic modes. Emerging out of the tradition of science and technology studies and influenced by sociologist Bruno Latour, ANT is a theoretical approach with a nondual, relational ontology. This means that rather than starting from a view that certain conceptual categories or variables matter and seeking to identify how they operate in a particular case, ANT starts from the particulars of the case and allows insights to emerge from the story.[22] Notably, in this approach, nonhumans are required to "do something and [not] just sit there"—meaning that they cannot be simply the material substrate upon which humans act (Latour 2005, 128, quoted in Robbins and Marks 2010).[23] Bakker and Bridge (2006) explore a number of approaches to the “re-materialization of human geography,” including work on commodities, corporeality, and hybridity. The challenge of the *material turn* that they pose is crucial to bear in mind throughout this book: “how to express the causal role of material without straying into object fetishism or without attributing intrinsic qualities to entities/categories whose boundaries are ‘extrinsic’—defined, at least in part, socio-culturally? How, in other words, can we insist on the importance of ‘things’ without treating them solely as things?” (Bakker and Bridge 2006, 14).

One can see what a difference one’s disciplinary and theoretical lenses make in the varied accounts we construct about urban nature. Looking at the urban forest, a city planner might argue that trees are part of the urban environment that contribute to resident quality of life. A Marxist approach might add that these trees have a role in attracting residents/labor to live/reproduce in that location. An approach guided by Marxism and ANT goes even further, as Perkins (2007) does, to argue that trees actively *labor* in this context. Similarly, looking at gardens, Power (2005, 48) demonstrates how planted species, pests, and weeds operate as subjects, “drawing people and plants into a relation of care” (quoted in Robbins and Marks 2010).[24] There are fertile examples of scholarship that bring the traditions of political ecology and ANT together, such as Paul Robbins’s (2007) investigation of the American lawn.

Much of the literature on urban politics and governance fails to sufficiently attend to the material realm. This book addresses that gap by continuing the project of “re-naturing urban theory” that urban political ecology scholars have begun and of “re-materializing” resource geography in an urban setting. It investigates the physical opportunities and constraints that the existing urban environment places on policymaking. Both biotic actors (trees, plants) and abiotic actors (buildings, sidewalks) are taken into account and investigated for their role in resource management practices. This book also examines links between the material and the discursive, by attending to what claims are made about the biophysical properties, benefits, and importance of nonhuman actors.

And vice versa, this work brings a detailed sense of urban politics into nature-society scholarship. Political ecology has a relatively longer tradition in rural settings and global south contexts than in urban areas and the global north and primarily builds upon a Marxist perspective (McCarthy 2002; Walker 2003; Robbins 2004). Therefore, political ecology does not always engage with the theories and traditions of urban politics and governance developed from global north contexts and outside of Marxist approaches. This text aims to help bring these literatures into productive conversation through a New York City case. In sum, this book is a critical examination of the practices of urban forestry and urban agriculture in a contemporary, North American, global city in the first decades of the twenty-first century. But it is critique couched in openness to possibility, interest in practice, and commitment to incremental change through the transformations of urban land and its uses.

Overview of the Book

The first section of the book provides historical context about New York City and then introduces PlaNYC. The governance of urban nature cannot be understood independent of the political-economic context in which it is embedded. Thus, chapter 1 presents a brief overview of how the changing political economy since the 1970s fiscal crisis has affected the management of the urban environment in New York City by both state and nonstate actors. In exploring the current environmental governance network in New York City, I present Social Network Analysis graphs of the urban forestry and urban agriculture organizational networks, which make visually apparent some of the key differences in the actors engaged across these issue areas. Chapter 2 examines the politics of PlaNYC agenda-setting, revealing the key actors involved in creating the plan, the basic assumptions driving the creation of the plan, and comparing and contrasting the treatment of forestry and agriculture in the plan.

The second section presents the urban forestry case. Chapter 3 describes the heart of the MillionTreesNYC campaign: planting one million trees. It focuses on the organizational, institutional, and material arrangements involved in transforming the city's urban forest. Chapter 4 goes beyond planting to explore the creation of an urban forestry movement through public outreach and engagement, the development of a network of advisors, and growth over time of the campaign.

The third section presents the urban agriculture case. Chapter 5 places the recent emergence of interest in urban agriculture and food systems in the context of the history of community gardening in New York City since 1970. It explores the vibrant material practices and varied narratives employed by practitioners engaging in urban agriculture from the 2000s to the present. Chapter 6 examines the ways in which civic and public actors working outside the boundaries of PlaNYC iteratively embedded urban agriculture and food policy into municipal planning efforts. It shows how the food systems framework was used to construct creative coalitions and articulate visions in order to influence the municipal agenda, including the 2011 update to PlaNYC.

The last section compares the cases and offers conclusions. Chapter 7 synthesizes these two cases, making thematic comparisons across the three research questions. It begins by returning to the central question of why urban forestry was so appealing that it merited its own signature mayoral initiative in PlaNYC, whereas urban agriculture was overlooked. Then, bearing in mind that the politics of urban nature is both framed by storylines and influenced by nonhuman actants, I widen this political analysis from a focus on politics and social networks to a focus on actor-networks and key narratives. Additionally, I observe how the cases shift over time in response to both internal and external factors, even within a four-year window.

Finally, chapter 8 closes with conclusions about how, why, and for whom we create urban forests and farms. Particularly, I explore these pieces of urban nature in the context of an early twenty-first-century global city that has engaged in municipal long-term sustainability planning, investment in green infrastructure, and creation of food policy visions. In some ways, there is a great degree of similarity across these two cases—such as the geographic context, historical moment, political leadership, and state of the economy. Yet this book reveals that even within a single city over a relatively narrow period of time, we find substantive differences in the urban forest assemblage and the urban agriculture assemblage.[25] Neither political-economic explanations nor traditional approaches to urban politics alone can explain the variation we observe across multiple axes. We must attend to humans and nonhumans, to language and to stuff, to hierarchies and to networks. As Timothy Mitchell (2002, 8) notes, "Theory lies in the complexity of the cases"; I offer the detailed cases and the synthesis across them as presentations of the processes involved in the construction of urban natures.

These cases from New York City help us understand and disentangle the political decisions, popular narratives, and physical practices that shape city greening. We can reflect on and raise questions about these processes in any city or town where decision-makers, residents, and activists are seeking to make transformations toward sustainability in the urban landscape and city institutions. New York is a global city currently wrestling with the challenges of sustainability and resilience that can serve as catalyst for other distant locales. Given current professional networks—such as the Urban Sustainability Directors' Network, Local Governments for Sustainability (ICLEI), and 100 Resilient Cities—policy ideas often circulate rapidly around the country and the globe (see also Peck and Theodore 2015). Understanding the processes at work *behind* these policy ideas provides insight into how and why these ideas came about and can inform future strategies. These complex stories are threaded through with power, agency, political choice, and moral questions. They are important to understand because how they play out influences the kind of city that we create, with very real consequences for the lives of urban dwellers.

1

GREENING NEW YORK CITY

Political Economic Context and Environmental Stewardship from 1970 to the Present

Following the political ecology traditions of "progressive contextualization" and tracing "chains of explanation" or "webs of relations," this chapter places the production of nature in New York City in a broader context (Blaikie and Brookfield 1987; Vayda and Walters 1999; Rocheleau 2008). While the long arc of New York City's political-economic history extends over centuries to include its precolonial, colonial, mercantile, and industrial periods (see, e.g., Burrows and Wallace 1999), the most crucial era to consider for its influence on contemporary forms of urban nature began in the 1970s. This period was characterized by postindustrial restructuring and transformation that has shaped New York's role as a global city in a globalized economy (Sassen 2001; Savitch and Kantor 2002; Sites 2003). Numerous scholars consider New York City's 1975 fiscal crisis as a political-economic turning point locally and globally, as part of the cyclically crisis-prone nature of capitalism (Shefter 1985; Sites 1997; Polanyi 1944 [2001]; Jessop 2002; Harvey 2005).

In studying this postindustrial shift, regulation theorists identified changes in urban governance, arguing that a transition from national, state-led government to multiscaled, networked governance occurred, involving both (1) an expansion of actors in the decision-making arena, and (2) shifts in the scales of those arenas (Harvey 1989; Jessop 2002). Regarding the expansion of actors, private sector and civil society actors play crucial roles in political decision-making, thereby changing the role of the state (Bulkeley 2005). The retreat of the state has been understood as a "hollowing out" (Jessop 1994, 2002) or as a "shadow state"—where civil society takes on state functions (Wolch 1990). Perkins (2009) argues, however, that the "shadow state" concept is less applicable to environmental management than it is to social services, for environment quality was never an entitlement in the way that welfare was, and it has always been an area of localized governance and state–civil society interaction.[1] Thus, it is clear that public, civic, and private actors are all shaping the urban sphere, but we can question whether the expansion of actors is anything new or whether the metaphor of hollowing out is completely fitting in the case of urban nature. Regarding shifting scales, there has been concurrent upscaling (as in the creation of the European Union) and downscaling (as in the rise of cities and regions in policy innovation) (Swyngedouw 2005; Gustavsson et al. 2009). Harvey (1989) argues that this shift has driven a change in the strategies of urban governance from "managerialism," focused on the provision of social services, to "entrepreneurialism," focused on interurban competition. Local sustainability initiatives like PlaNYC can thus be understood in the context of competitive, global cities engaging in city image-making through sustainability planning and investments in environmental quality (While et al. 2004; Gibbs and Krueger 2007; Jonas and While 2007). While examining the practices of local politics is crucial to understanding municipal sustainability planning, scholars remind us not to privilege a single scale and to place local phenomena in broader temporal and spatial contexts, which I do in this chapter by drawing upon secondary literature on the evolution of the governance of urban nature in New York since 1970.

In addition to this brief historical review, my own approach to understanding the governance of New York's urban environment draws upon survey results, Social Network Analysis (SNA) data, and key informant interviews.[2] SNA is a quantitative method rooted in graph theory that provides a way to analyze complex networks (Wasserman and Faust 1994).[3] Using SNA to examine local "ego networks," one can begin to understand the relations between actors and the relative positioning of these actors in their local networks (Wellman 1979; Marsden 1990; Scott 2000; Burt 2007; Connolly et al. 2013). In New York City, there is a large, polycentric network of civic groups working on a range of site types across differing neighborhoods helping to steward urban nature (Connolly et al. 2014). A subset of these civic organizations is playing a crucial brokering role in the network, helping to share resources and information across sectors and scales (Connolly et al. 2013). Indeed, most of these broker groups were founded in the post-1970s era and persist to the present day. Finally, with this political-economic and civic stewardship network data as background, we can then examine the contemporary governance networks involved in urban forestry and agriculture, which provides an overview of the actors and relationships described in subsequent case chapters.

Political Economic History: New York City's Urban Nature from 1970 to the Present

It is important to note that many of the federal policies that helped encourage the suburbanization of the country and the form of New York City's open spaces precede the fiscal crisis and date to the Depression and prewar years.[4] So, too, did the Robert Moses era of top-down, centralized, car-dependent planning and road building leave an indelible mark on the New York City landscape. Moses' legacy includes many of the parks and beaches in the city and region—as well as the parkways, highways, and bridges connecting these sites (Caro 1975; Gandy 2002). Brecher et al. (1993, 13) attribute many of DPR's ongoing budgetary challenges to the "mismatch" between the size of the park system, expanded under Moses with substantial federal support, and the limited local resources available to maintain it since that time.

From the 1970s, New York City faced declines in the manufacturing base, the rise of the service economy, a rapid population outflow to suburban areas (particularly among whites), a decline in the municipal tax base, and widespread housing abandonment and arson (Mollenkopf and Castells 1992; Sullivan 1992; Brecher et al. 1993; Gandy 2002; Harvey 2005; Berg 2007). Even with substantial continued immigration to the city, the overall city population declined from 1970 to 1980 and did not rebound until the real estate and development boom of the 1990s (NYC DCP 2011). These changes fundamentally reoriented the spatial organization, demographic patterns, and social structure of the city. Table 1.1 shows New York City's total population and racial and ethnic composition over the period 1970–2010.[5]

In the 1970s, sustained municipal budget deficits coupled with borrowing for municipal operations nearly led the city to default on its loans. This near-bankruptcy had major consequences for city governance through the creation of new public authorities that were not responsible to the electorate (Shefter 1985). Shefter's (1985) detailed analysis of the fiscal crisis reveals that its origins were as much political as economic; for mayoral candidates to succeed in being elected, they had to put together grand, multiracial, multiethnic, cross-class coalitions that supported municipal spending on public wages, services, and capital projects. Postcrisis, when the federal government refused to offer New York City a key bailout (although it later acquiesced), the *New York Post* headline "Ford to City: Drop Dead" epitomized the sense that the city was being left to its own failures (Shefter 1985; Roberts 2006). Overall, Brecher et al. (1993, 10) identify four

TABLE 1.1 Population of New York City (1970–2010) by race and ethnicity

YEAR	TOTAL	WHITE	WHITE (NON-HISPANIC)	BLACK	BLACK (NON-HISPANIC)	OTHER	HISPANIC (OF ANY RACE)
1970*	7,894,862	6,048,841 (76.6%)	4 969 749 (62.9%)	1,668,115 (21.1%)	N/A	138,236 (1.8%)	1,278,630 (16.2%)
1980*	7,071,639	4,294,075 (60.7%)	3,668,945 (51.9%)	1,784,337 (25.2%)	N/A	749,902 (10.6%)	1,406,024 (19.9%)
1990*	7,322,564	3,827,088 (52.3%)	3,163,125 (43.2%)	2,102,512 (28.7%)	1,847,049 (25.2%)	852,714 (11.6%)	1,783,511 (24.4%)
2000+	8,008,278	3,806,508 (47.5%)	2,801,267 (35%)	2,274,049 (28.4%)	1,962,154 (24.5%)	1,084,303 (13.5%)	2,160,554 (25%)
2010#	8,175,133	3,797,402 (46.5%)	2,722,904 (33.3%)	2,228,145 (27.3%)	1,861,295 (22.8%)	1,254,858 (15.3%)	2,336,076 (28.6%)

N/A—not available

Sources: *Gibson, Campbell, and Kay Jung. 2005. "Historical Census Statistics On Population Totals By Race, 1790 to 1990, and By Hispanic Origin, 1970 to 1990, For Large Cities And Other Urban Places In The United States." Washington, D.C.: U.S. Census Bureau, Population Division. Working Paper No. 76. February 2005. Accessed online via: http://www.census.gov/population/www/documentation/twps0076/twps0076.html (17 June 2013)

+ US Census Bureau. 2000 Census. SF-1, Matrices P7 and P9. Accessed online via American Fact Finder: http://factfinder2.census.gov/faces/tableservices/jsf/pages/productview.xhtml?pid=DEC_00_SF1_QTP5&prodType=table (17 June 2013).

US Census Bureau. 2010 Census. DP-1: Profile of General Population and Housing Characteristics. Accessed online via American Fact Finder: http://factfinder2.census.gov/faces/tableservices/jsf/pages/productview.xhtml?pid=DEC_10_DP_DPDP1 (17 June 2013).

periods of distinct policy and budgetary regimes in New York City from 1960 to 1990 and three main factors contributing to these changes: "the performance of the local economy, intergovernmental interventions, and power relations among local interest groups."[6]

While other urban scholars have focused on housing policy and local development practices (Sites 1997), I examine the ways in which the economic downturn influenced the production of nature in New York City. One commonly cited consequence for open space management was the way in which massive budget cuts led directly to declines in the staffing, maintenance, and safety of New York City's parks, including sustained challenges with addressing graffiti and vandalism throughout the 1970s and early 1980s (Brecher et al. 1993). Most visibly, planned capital investments in Central Park that were called for in a 1973 master plan were postponed (NYC DPR 2013a). In 1974, as a strategy to ease the park maintenance burden, the city transferred more than thirteen thousand acres of land around Jamaica Bay and the Staten Island coast to the National Park Service, land that now comprises the federal Gateway National Recreation Area. Overall, the New York City Department of Parks and Recreation's (DPR) operating expenditures underwent a fairly steady decline from 1975 to 1984, with an increase in the budget in the second half of the 1980s (Brecher et al. 1993). Gandy (2002, 104) describes the impact of such long-term declines in municipal funding, leading to vastly unequal access to public space throughout the city: "By the early 1990s New York ranked nineteenth among major American cities in terms of per capita public expenditures on its park system (far below Los Angeles or Chicago, for example) and was left with just half the park staff it had in 1960."

At the same time, economic decline and housing abandonment also triggered a rise in community gardening practices (Francis et al. 1984; Lawson 2005). Many of the thousands of vacant lots across the New York City landscape of the 1970s and 1980s—particularly in the South Bronx, Lower East Side, and Bushwick—were the result of arson (Sullivan 1992). Intrepid residents worked to reclaim vacant lots, fight back against drug sales and prostitution, and create neighborhood green spaces. The more than five hundred gardens currently in New York City display a spatial distribution informed by the disinvestment of the 1970s (see Map 1.1). The individual efforts of gardeners are well-documented within the contemporary discourse of community gardens. What is perhaps less explored are the network of civic organizations and institutions that emerged out of this period to support gardening and greening efforts, which are further discussed in the section on civil society.

The 1980s and early 1990s saw the continued rise in inequality across New York City, with the metaphor of the "dual city" encapsulating the differences

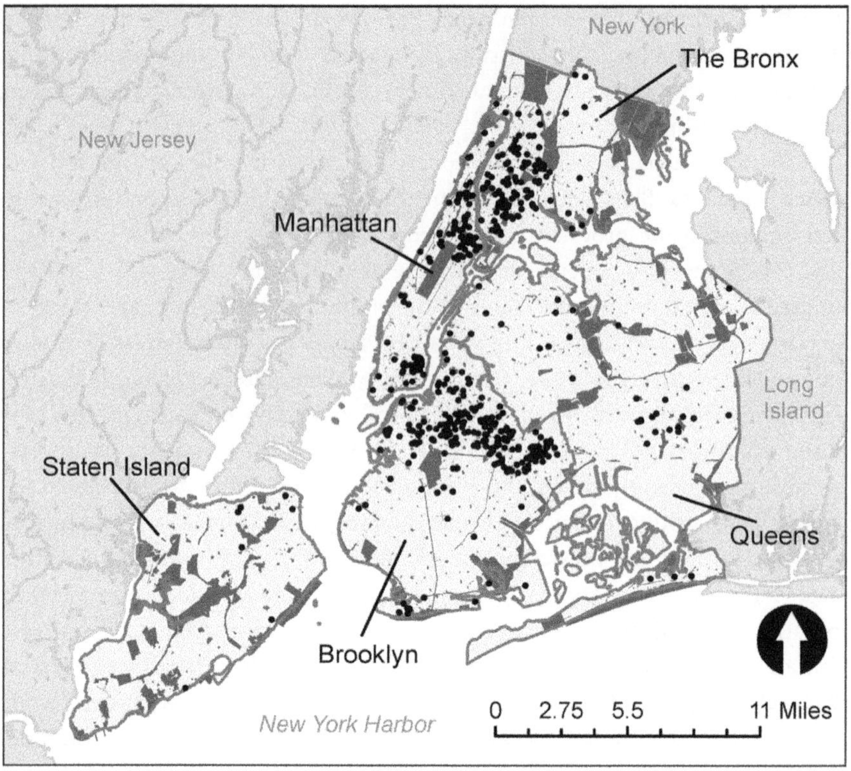

MAP 1.1 Map of New York City's parks and community gardens. Created by: Michelle L. Johnson, U.S. Forest Service. Data sources: Greenthumb. 2013. New York, NY: New York City Department of Parks & Recreation. Available via: https://data.cityofnewyork.us/Environment/Greenthumb/86sd-4yhi. (September 10, 2013); Parks. 2013. New York, NY: New York City Department of Parks & Recreation. Available via: https://data.cityofnewyork.us/Housing-Development/Map-of-Parks/jc79-4imn. (September 10, 2013); U.S. GDT Federal Park Landmarks. 2002. Redlands, CA: ESRI and Geographic Data Technology, Inc.; New York City State Parks. City of New York, NY.

between Wall Street, yuppie elite, and the low-income communities of color ravaged by disinvestment, crack cocaine, and AIDS. This economic period of the dual city saw the political transition from Mayor Ed Koch (1978–1989) to the first African American mayor of New York City, David Dinkins (1990–1993) (Mollenkopf and Castells 1992). In 1989, the rape and brutalizing of a white woman in Central Park, allegedly by a group of African American youths (who were exonerated decades later by DNA evidence), came to serve as an incredibly racially divisive moment in the city—and furthered the sense that New York City's parks were unsafe (Knight-Ridder newspapers 1989; Mollenkopf 1992;

Filipovic 2012).[7] At the same time, the city was able to increase its capital spending on parks and open space, starting in the 1980s. According to a history of the NYC DPR: "As the city reentered the municipal bond market in 1981, Mayor Koch issued his first ten-year capital plan. The plan proposed a $750 million commitment to rebuild the city's parks. For the first time in years, the Parks Department was also building up its permanent work force, which had fallen to under 2,500 workers in 1980 from over 5,200 in 1965" (NYC DPR 2013a). This mayoral and agency commitment was mirrored by the support of local community boards for parks and open space, which made park maintenance their number one priority—surpassing police patrols for the first time in 1986 (NYC DPR 2013a). The spatial inequalities in the way that parks were managed also helped to catalyze the local environmental justice movement in New York City to focus on access to quality open space as a crucial area of concern (Francis et al. 1984; Fox et al. 1985).

The development boom of the mid- to late 1990s was coupled with and encouraged by the Republican mayor Rudolph Giuliani (1994–2001) (Sites 1997; Berg 2007). Giuliani emphasized aggressive enforcement of "quality of life" violations (loitering, drunk and disorderly, jay walking, graffiti) in ways that were praised by some for reducing crime but also criticized as anti-homeless, intolerant, and racist (Smith 1998; Grogan and Proscio 2000). Inspired by James Q. Wilson's "broken window theory," Giuliani argued that these smaller violations led to a disordered public sphere that signaled a lack of caring and invited further, more serious crimes (Wilson and Kelling 1982). This administration was notoriously prodevelopment, with Giuliani criticizing community gardeners as "stuck in the era of communism" (Lefer 1999). When the mayor attempted to auction off several hundred garden sites for housing development, this triggered a community garden crisis that is described in chapter 5. In this period, New York's urban nature was also shaped by Henry Stern, the eccentric commissioner of the DPR from 1983 to 1990 and 1994 to 2000. A 1995 *New York Times* profile described Stern's leadership under Giuliani: "Mr. Stern . . . presides over a Parks Department that has been under siege for 25 years. He battles daily to save an empire of 1,500 properties and more than two million trees from further budget cuts at a time when layoffs will soon reduce his roster of full-time workers to a record low of 2,400. He had twice that many when he was Parks Commissioner in the last decade. . . . Today, under a Republican mayor, he must manage the citizens groups and donors who help the parks in lean times" (Bumiller 1995). Overall, Giuliani established a revanchist form of governance that reduced taxes, cut public services, and heavily policed the streets of New York, serving certain citizens (white, wealthy), while ostracizing and criminalizing others (homeless, poor), and transforming the public realm in the process (Smith 1998; Harvey 2005; Berg 2007).

From a critical perspective, Bloomberg-era efforts at local sustainability planning could be understood as more sophisticated and subtle modes of state engagement in the perpetuation of capital accumulation. Jessop (2002) argues that postindustrial contemporary regimes emphasize the role of competition, the entrepreneur, innovation, scales other than the nation, and sectors beyond the state. This pattern is evidenced in New York City's economy, which has seen a precipitous decline in manufacturing and a rise in services, entertainment, finance, and advertising (Sassen 1991; Harvey 2005; Berg 2007). With the footloose nature of both capital and labor, the city works to attract residents and businesses that it deems as desirable (e.g., educated, wealthy, "creative class"). One of the ways that government can attract people is by providing amenities, which include trees, open space, and green infrastructure (as well as other services not examined here, such as schools). In keeping with Jessop's observation of rescaling and postnationalism, municipalities are developing these plans and green infrastructure campaigns as they compete amongst themselves to create desirable, livable places. The rise of public-private partnerships and multisector advisory boards, discussed below, also testifies to the shift from states to regimes. These institutional structures allow for crossing land jurisdictions, raising private funding, and leveraging the expertise and inputs of a wide network of actors.

As noted in the text of PlaNYC itself, economic and demographic conditions were extremely different in 2007 from thirty years prior in the mid-1970s, allowing for a substantial increase in municipal government spending. In spring 2007, the city budget was in surplus—as it had been for the prior four years; population and in-migration were booming; and the national economy was strong (until the fourth quarter of 2007) (City of New York 2007; ICLEI 2010b). PlaNYC sought to sustain that growth in a way that ensured the livability and economic competitiveness of New York City by committing millions of dollars in municipal capital investments in infrastructure, housing, transportation, and open space (City of New York 2007). Thus, we can examine PlaNYC as the latest form of growth-promotion of the entrepreneurial urban regime—albeit one that has more thoughtful attention to environmental quality and public health than prior schemes, as will be further discussed in chapter 2.

Civil Society Involvement in Environmental Stewardship since 1970

A wave of *civic environmental groups* emerged starting in the 1970s in response to both the local experience of the urban fiscal crisis just discussed, as well as in the context of the rise in environmental consciousness associated with national environmental legislation, such as the Clean Water Act, the Clean Air Act, and the

Endangered Species Act (see Mertig et al. 2001). For example, the organization Green Guerillas was founded by Liz Christy in 1973 (and later incorporated in 1976). Initially, Christy used "seed bombs" to re-green vacant lots in Manhattan's East Village and Lower East Side and later founded the first community garden in that area. The Green Guerillas were later instrumental in community organizing during the garden crisis of the 1990s, and they remain an active and critical node in the citywide network of community gardens. Similarly, in 1975, Trees New York was founded as a consortium of groups in direct response to the city's inability to care for its urban forest. This consortium later was incorporated as a nonprofit in 1982 and persists today as one of the civic environmental umbrella groups focused on the urban forest. One nonprofit leader recalled that its organization's founders put an ad in the *New York Times* to solicit community volunteers and received five thousand responses the first day. He cited as motivation "the idea that jobs held by City employees who lost their jobs would be filled by ordinary citizens. Obviously not police or fire or medical people, but very simple stuff like caring for street trees or serving meals in a senior center." Such jobs included citizen-managed garbage pickups at central locations to compensate for reduced sanitation runs (respondent 57). This era of rising environmental awareness was also when the Park Slope Food Cooperative—a member-owned grocery store that currently has more than fifteen thousand members—was formed (NYC Council 2010). Thus, the fiscal crisis affected not only the material form of urban open spaces but also the organizational landscape of groups that engage in these issues.

Throughout the 1970s in New York City, there were diverse examples of civil society groups, municipal leaders, and local elites developing *collaborative governance efforts*. First, in 1970, Mayor John Lindsay approached Marian Sulzberger Heiskell (of the Heiskell family, owners of the *New York Times*) about forming a public-private partnership to support New York City's local environment. Out of this initial concept, the Council on the Environment of New York City (CENYC, which changed its name to GrowNYC) was formed by executive order of Mayor Lindsay. This hybrid entity describes itself as "a privately funded citizens organization in the Office of the Mayor" and has a board that includes both private individuals as well as appointees of the mayor (GrowNYC 2012). In 1974–1975, CENYC created its Open Space Greening Program and hired Liz Christy to help develop that program.

Civic and government environmental programs continued to proliferate and mature in the 1970s and 1980s. By 1976, CENYC began its Greenmarket program of farmers markets, with the Union Square Greenmarket being one of the earliest participating sites. GreenThumb, the municipal program to support the burgeoning community garden movement, was also created in 1978. Responding to a perceived need for coordination amongst the many neighborhood-based and citywide greening organizations in New York City, the Neighborhood Open Space Coalition was founded in 1980 (Lawson 2005). Some organizations underwent a process of maturation, from coordinating individual volunteers to working as a node that served small community groups. One respondent recalled:

> And then in the '80s, maybe five years later, we began focusing on low income neighborhoods, which is where we are now. So we provide small cash grants to neighborhood groups beginning between $500 and $3,000. Sometimes larger. Sometimes up to $10,000. These are generally not 501(c)(3)s. They can't have any paid employees. They're civic associations, tenants' organizations, they're gardening clubs. There are a lot of block associations. And sometimes they're just people who want to get together and deal with the problem; they aren't even an organized group. . . . (respondent 57)

Finally, building upon the momentum of these locally rooted efforts, national programs began to emerge to support the community-gardening movement. For example, the American Community Gardening Association was founded nationally in 1978, and the USDA Cooperative Extension Urban Gardening Program was operating in twenty-three cities as of 1976 (Lawson 2005; McMillan 2010).

During the 1980s and early 1990s, in response to the declining quality and safety of parks, these sites were increasingly managed through *public-private partnerships*. Mayor Koch took the first steps toward privatization of certain park functions by bringing in contracted concessioners to operate facilities and by transferring the operation of zoos to the New York Zoological Society (now the Wildlife Conservation Society) (Brecher et al. 1993). The public-private partnerships of the Central Park Conservancy (formed in 1980) and Prospect Park Alliance (formed in 1987)—as well as more recent efforts like the Friends of the High Line (formed in 1999)—are all attempts to harness private support for parks in the face of shrinking municipal budgets (Gandy 2002; Svendsen 2010; City of New York 2012a). Critics note, however, that these partnerships, which are not evenly distributed across all New York City parks citywide, exacerbate a "two-tiered" park system, whereby wealthy neighborhoods can afford to form "friends of" groups that can enhance scant public spending, whereas poorer neighborhoods cannot (Harden 1999; D. Taylor 2009; Svendsen 2010).

Continuing to the present day, several public-private partnerships support open space and the urban environment *citywide*, rather than at specific park sites. These include the New York Tree Trust (formed in 1994) and Partnerships for Parks (formed in 1995), both of which are nonprofit entities incubated directly out of DPR. The former helps support the maintenance, care, and planting of trees, while the latter helps incubate local, community-based groups to bring private and community support to local parks. The Historic House Trust (formed in 1989) is a private nonprofit that helps support and manage historic homes and museums on DPR land. Most recently, in 2012, DPR incubated the private nonprofit Natural Areas Conservancy to support the city's "natural area" parks, with initial private funding from the Tiffany & Co. Foundation and the Doris Duke Charitable Foundation (NYC DPR 2012).

A recent study helps to understand the size, scope, and composition of the contemporary *civic environmental stewardship network* in New York City. Researchers surveyed approximately three thousand civic groups to identify their organizational foci, geographic turf, and social networks with government, business, and civil society (Fisher et al. 2012; Connolly et al. 2013, 2014; Svendsen et al. 2016). Stewardship is defined as doing any of the following: conserving, managing, monitoring, advocating for, or educating the public about local land, air, water, waste, energy, or toxics issues.[8] Map 1.2 shows the spatial distribution of these stewardship groups, with clustering in Central Brooklyn, the South Bronx,

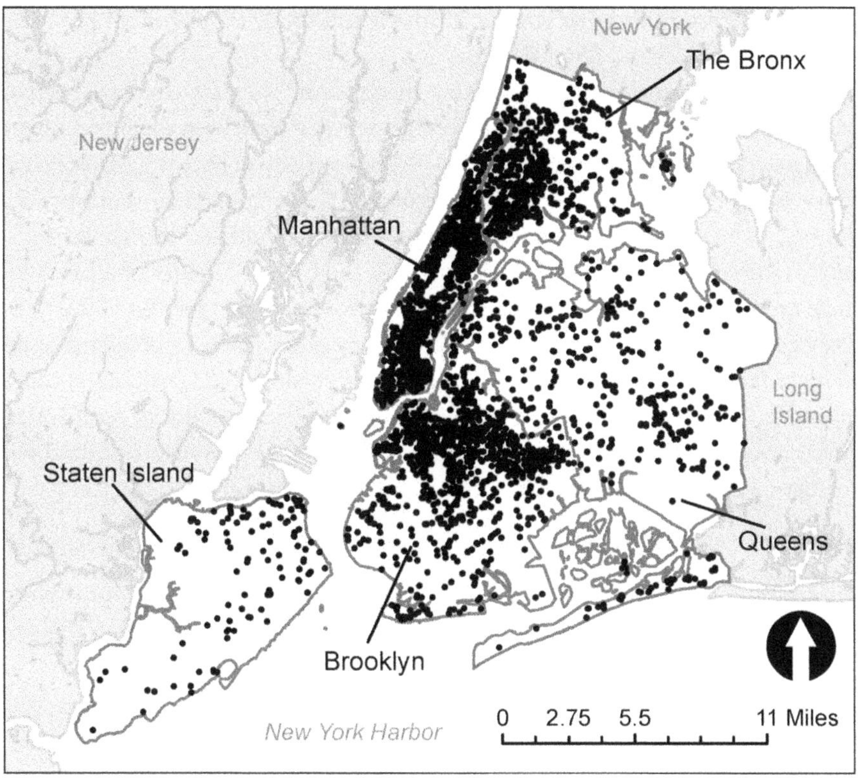

MAP 1.2 Map of office locations of civic stewardship groups in New York City. Created by Michelle L. Johnson, U.S. Forest Service. Data source: STEW-MAP database. U.S. Forest Service 2007.

and throughout Manhattan and a lower density of groups in more suburban areas of Queens and Staten Island. The physical site types on which these groups work are quite diverse, with the most frequently stewarded sites including parks, street trees, and community gardens.[9]

The civic stewardship groups range widely in size and formality from multimillion-dollar environmental nonprofits to completely grassroots, non-501(c)(3) community gardens, clubs, and block associations. Overall, the degree of professionalization of these groups is rather low. The majority (58 percent) have either zero or one staff member working at the organization, and 54.2 percent have a budget of less than $10,000. At the same time, 19.7 percent of groups have eleven or more paid staff members, and 11.6 percent have an annual budget greater than $1 million. Groups that work on green buildings are more likely to be more highly professionalized, whereas groups that work on community gardens are more likely to score low on a professionalization index that takes into account size of budget and number of paid staff (Fisher et al. 2012). This squares with the understanding of community gardening as a grassroots movement, whereas green buildings are created through real estate development practices requiring technical and financial expertise.

Finally, these groups are not working in isolation. They are connected across sectors and scales—often via a handful of groups that have been identified as key "bridging organizations" or brokers via SNA (Connolly et al. 2013). These civic stewardship brokers channel external resources (from government and private donors) to help coordinate and support the work of smaller-scale organizations, such as community gardens, block associations, and friends-of-parks groups. Often these groups have expertise in a particular focal area or site type (See Table 1.2). They have a "bi-modal" relationship with the state, consisting of cooperation in some instances (joint projects, public-private partnerships) and criticism in other instances (advocacy, litigation). Overall, these groups are building multiscaled governance capacity and helping build a more flexible, adaptive, institutional structure for the management of urban nature (Connolly et al. 2013, 2014).

TABLE 1.2 New York City civic environmental stewardship broker organizations as of 2007

ORGANIZATION NAME	FOCUS AREA
Green Guerrillas	Community gardens
GrowNYC (formerly CENYC)	Neighborhood greening, community gardens, farmers markets, waste reduction, recycling, environmental education, school gardens
Brooklyn Botanic Garden	Botanical garden, horticultural education
Just Food	Food and agriculture, farmers markets, CSAs, education

(Continued)

TABLE 1.2 Continued

ORGANIZATION NAME	FOCUS AREA
Trust for Public Land	Open space conservation, community garden preservation, schoolyards-to-playyards initiative
New York Cares	Volunteerism
New York Restoration Project	Cleaning and greening: Northern Manhattan parks, community gardens, MillionTreesNYC
Trees New York	Urban forestry, environmental education
American Littoral Society	Waterfront preservation, conservation of marine life
Citizens Committee for New York City	Supporting neighborhood engagement and activism, particularly in low-income communities
Park Slope Civic Council	Neighborhood preservation and quality-of-life advocacy in Park Slope, Brooklyn
Bronx Land Trust	Bronx community gardens
Municipal Arts Society	Creating a more livable city: design, architecture, planning, preservation

Source: Connolly, James J., Svendsen, Erika S., Fisher, Dana R., and Campbell, Lindsay K. 2013. "Organizing Urban Ecosystem Services through Environmental Stewardship Governance in New York City." *Landscape and Urban Planning* 109: 76–84, with author's description of focus areas added.

Visualizing Urban Forestry and Urban Agriculture Networks in New York City

In addition to examining the entire civic stewardship network in New York City, SNA allows for a visual means of exploring the differences in the networks of urban forestry and urban agriculture. Visualizations reveal the more centralized, state-led network of urban forestry in contrast to the more diffuse, polycentric network of urban agriculture. I generated diagrams for the forestry network (based upon thirty-four respondents in fourteen planning, forestry, natural resource, and environmental organizations) and the agriculture network (based on forty-three respondents in thirty-six planning, gardening, agriculture, open space, and food system organizations). Dots represent organizational nodes, and lines represent collaborative ties between them.[10] The size of the node reflects the number of ties, both the number of groups that identified working with that group and the number of groups identified as partners by that organization. Government groups are shown in black; civic groups are shown in light gray; private sector groups are shown in medium gray.

The forestry network is centered upon DPR as the agency in charge of the management of trees in the city. I present two different views of the forestry network: one that distinguishes between different DPR programs and divisions—leadership, MillionTreesNYC, Partnerships for Parks, Central Forestry and Horticulture (CFH), Natural Resource Group (NRG), and GreenThumb (figure 1.1); and one that combines them all into a single node (figure 1.2). Figure 1.2 most clearly demonstrates the prominence of this agency within the network, with a far greater number of ties than any other node in the network. Even the nonprofit half of the MillionTreesNYC partnership, NYRP, is relatively smaller than the DPR node and is similar in number of ties to some of the main civic groups in the MillionTreesNYC Stewardship Corps, including Trees New York and the botanical gardens. This representation shows the importance of DPR to the forestry case, which the narrative in subsequent chapters will show extended from the earliest stages of PlaNYC goal-setting throughout all the stages of implementation. In figure 1.1, with the programs and divisions of DPR presented separately, we see a slightly more complex picture, wherein each of these divisions has its own, relatively distinct set of partner groups with which it works. In this representation, one sees that NYRP and DPR's MillionTreesNYC divisions have roughly the same number of ties. In both views, we see that there are just a few business groups with more than one tie, including DPR's contractors, nurseries, and some of the key corporate funders of MillionTreesNYC (Toyota, BNP Paribas, Home Depot).

The agriculture network immediately presents quite a different visual display that reflects its history as a civic-led movement as well as its recent turn toward entrepreneurial endeavors (see figure 1.3). Both the number of nodes and the number of connections between them is greater than in the forestry network. The central component of the network is roughly divided between civic and government groups. There are more private sector actors in the agriculture network than there are in the forestry network due to the entrepreneurial ventures in urban farming, farmers markets, and restaurants selling local produce that are proliferating in the city. Prominent civic groups include professionalized nonprofits (GrowNYC, Just Food, the botanical gardens, Trust for Public Land, Citizens Committee), grassroots-oriented or neighborhood-based groups (Green Guerillas, Added Value, East New York Farms), and research and funding entities (Columbia University, Cornell University, Doris Duke, Heifer International). The government groups have a fair amount of overlap with the forestry case (DPR, Office of Long-Term Planning and Sustainability [OLTPS], and New York City Housing Authority [NYCHA]) as there are a limited set of entities focusing on land management in the city. This network also shows the role of the New York State Department of Agriculture and Markets, former Manhattan borough president Scott Stringer's Office, and various city council members, through the planning efforts that are described in chapter 6.

Comparing these diagrams supports the claim that agriculture is a sprawling set of overlapping and distinct coalitions, whereas forestry is a more centralized

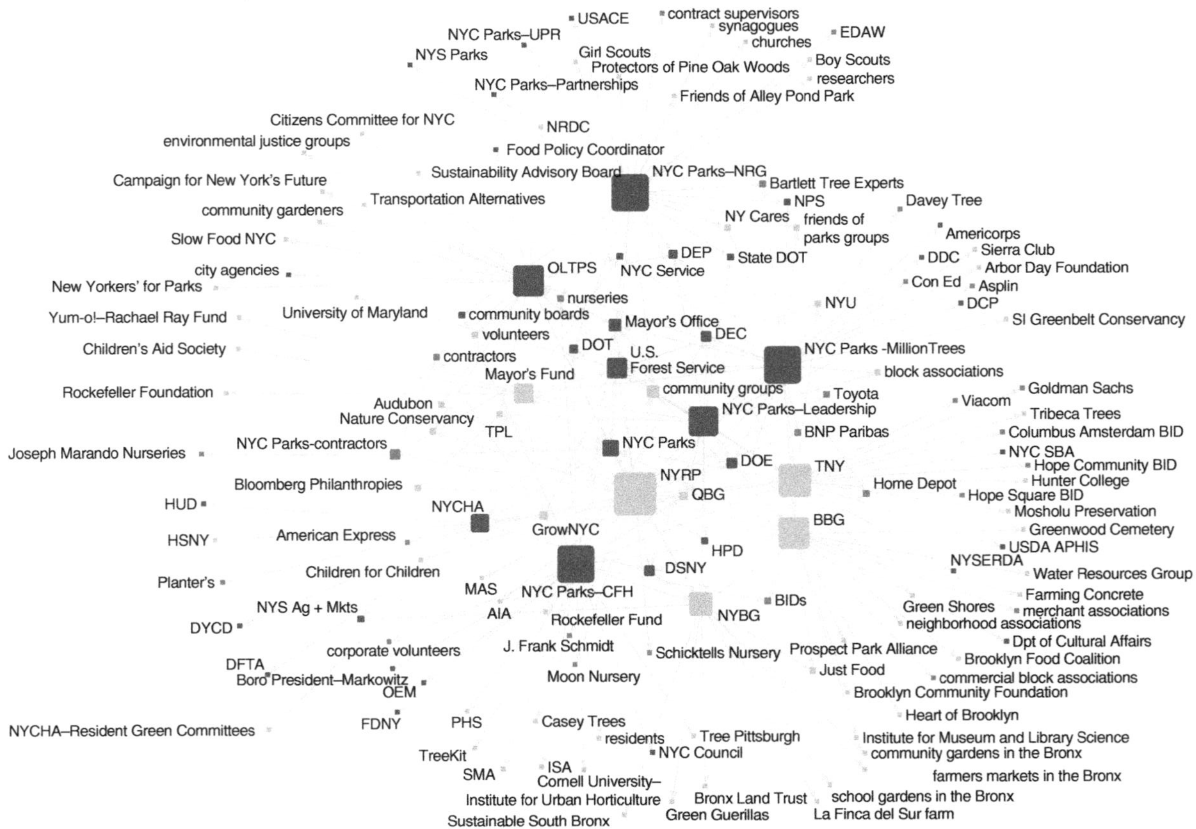

FIGURE 1.1 Urban forestry network diagram with DPR divisions disaggregated.

Source: Diagram created by author. Illustration assistance from Isaac Gertman.

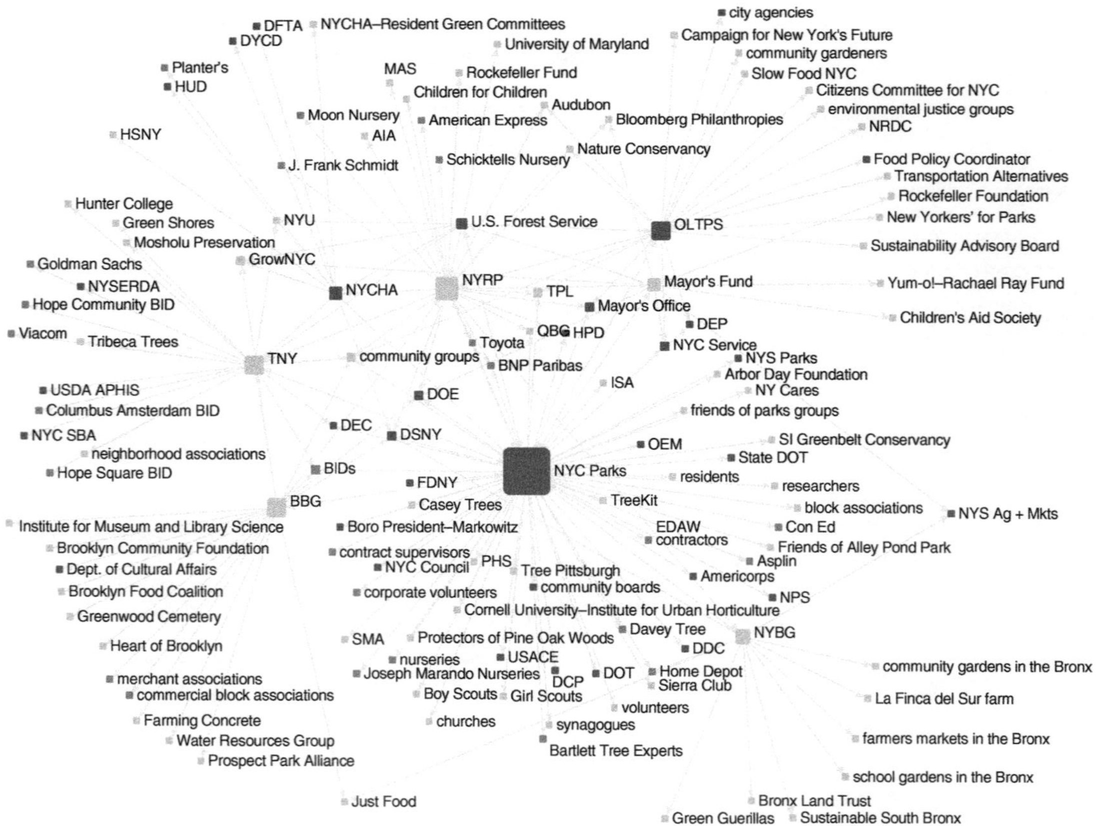

FIGURE 1.2 Urban forestry network diagram with DPR divisions combined.

Source: Diagram created by author. Illustration assistance from Isaac Gertman.

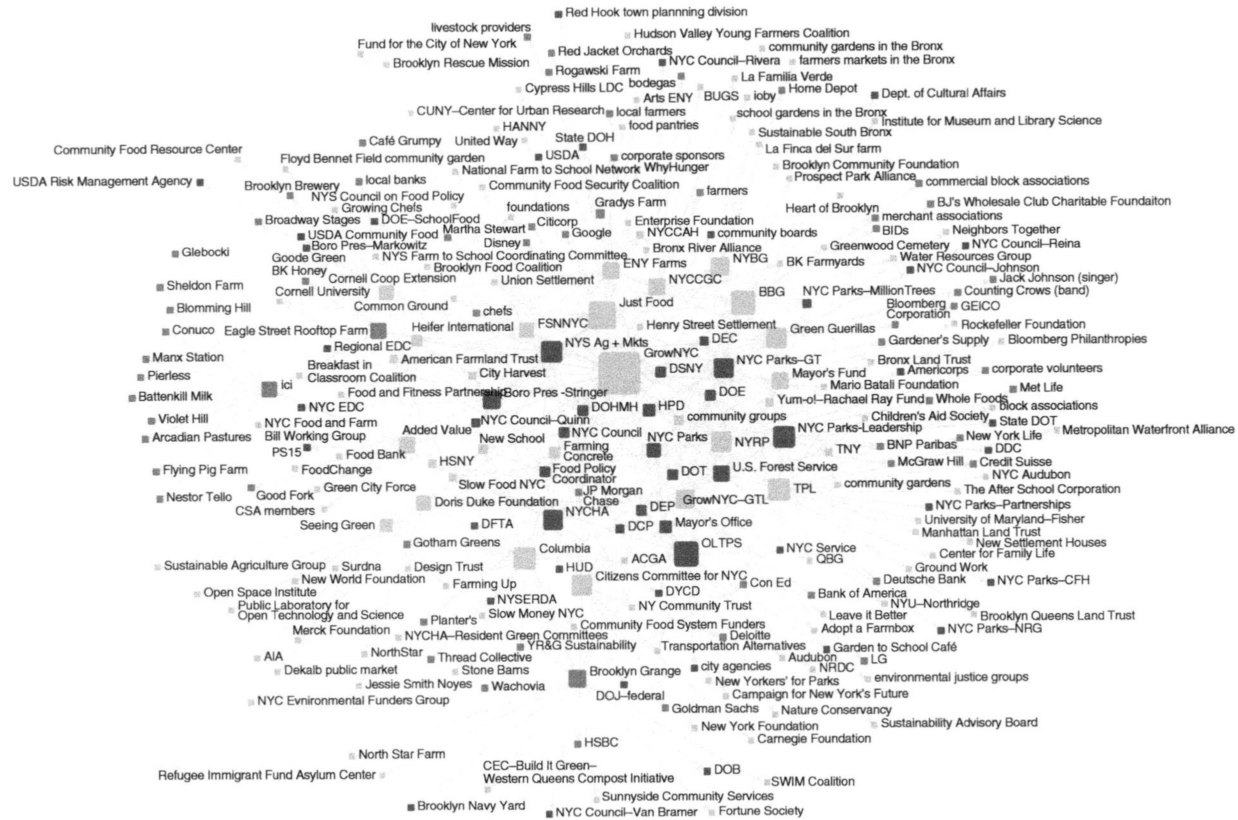

FIGURE 1.3 Urban agriculture network diagram.

Source: Diagram created by author. Illustration assistance from Isaac Gertman.

effort that is primarily led by DPR. Both networks are broad and include a wide range of actors, but the agriculture network appears more complex. I'll briefly review the role of the different types of actors shown in each diagram here.

- Both cases include *global corporations,* which serve as funders, donors, and sources of volunteer labor; New York City is unique in the preponderance of global firms that are located there and may choose to engage in corporate social responsibility through volunteer events "in their own backyard."
- Both networks include *city agencies* that control land and resources and regulate the urban environment. In the case of forestry, DPR is prominent, with a high degree of expertise in managing trees and a depth of resources devoted to sustaining the urban forest. DPR also has responsibility for many of the gardens in New York City, via the GreenThumb program, but it devotes markedly fewer staff and financial resources toward this end than it does to maintaining the city's urban forest.
- We can also see the influence of *elite actors* in both networks, although this is somewhat obscured in an organizational-level image of nodes. Most notably, Bette Midler's celebrity provided her with access to political leaders and donors and visibility for the endeavors of her nonprofit, NYRP, as will be further described in chapter 3. In the agriculture case, we see the role of celebrity chefs-as-donors, such as Rachael Ray and Mario Batali, as well as a few musicians, like Jack Johnson and Counting Crows, who donate to community-gardening efforts.
- The largest civic nodes in both cases are generally *professionalized, citywide nonprofits* that serve as brokers, providing information, resources, and services to their constituents, and bridging to organizations in other sectors and fields (see also Connolly et al. 2013).
- Finally, we see a diverse array of *grassroots or community-based informal groups* as well as *small businesses and business organizations.* It is important to note that many of these groups are not fully enumerated or named in the diagrams because large categories of groups—like block associations, community gardens, and volunteer groups—are used as placeholders for the hundreds of small groups in these categories.

Although the diagrams are not representations of the total network, these are the core organizational actors involved in each of the cases, during the period 2007–2011. Networks change—and these depictions represent a moment in time for these fast-evolving fields. Further nuance could be brought by examining the types of ties that link actors together (e.g., the flows through the network) as well as the role of particular individuals (in addition to organizations) in the network. These diagrams serve as visual representations that can be used alongside the case narratives presented in chapters 3 through 6 in order to track key nodes, peripheral nodes, and the connections among them. The centralized, professionalized, and state-led forestry network looks strikingly different from the more polycentric, civic-led agriculture network. Bearing in mind the distinct structures of these two networks, how do the various actors in these networks interact to influence the sustainability agenda and construct urban nature?

2

CREATING PLANYC

The Politics of Urban Sustainability Planning

Municipal decision-making can be opaque. While legislation and budget changes have formalized processes for public input, high-level strategic decisions are often made behind closed doors, with public comment sessions held after the fact.[1] While PlaNYC was praised for its ambitious scope, its breaking of bureaucratic silos, and its long-term vision, critics claim that PlaNYC's creation was top-down, technocratic, and antidemocratic. Indeed, PlaNYC was not a traditional plan held to the same processes of public planning that guide, for example, the Department of City Planning. Despite the name, it was a set of strategic initiatives originating in the mayor's office at City Hall.

This chapter examines the actors and processes involved in the development of the 2007 sustainability plan.[2] Certainly, PlaNYC was first and foremost a Bloomberg initiative. But the mayor did not act unilaterally; city agency leaders helped shape goals that would ultimately be the agencies' job to implement. The City Charter and administrative code establish the roles of the mayor and the agencies but allow room to maneuver based on individual leaders' interests. One layer removed in influence was the network of civil society groups engaged in the stewardship of New York City's environment that were just described in chapter 1. Even though they were not often consulted in the plan's conceptualization, the longstanding efforts of civil society groups inspired or informed the plan's goals.[3] One layer further removed from the process was the public—which was invited only at a late stage to comment on ten goals that were already well solidified. Held under the mantle of public consultations, these meetings instead allowed interested parties to feel included and helped get the word out about the PlaNYC agenda. Finally, juxtaposing the treatment of urban forestry and agriculture throughout the development of the first edition of PlaNYC sheds light on processes of framing and goal setting.

The Mayor: Bloomberg's Green Growth Agenda

The origins of PlaNYC are tied to the Bloomberg administration's (2002–2013) economic development policies, rezoning efforts, and a desire to accommodate a growing city population. Some of the long-term thinking about New York City land use and infrastructure was initially developed by Dan Doctoroff, who founded the nonprofit NYC2012 to head New York's bid to host the 2012 Olympics. Doctoroff, a former private equity fund manager, later became deputy mayor for economic development under the Bloomberg administration from 2001 to 2007 and is currently CEO of Bloomberg LP—a financial information company (ICLEI 2010b). The (failed) Olympic bid was a quintessential example of the sort of localized, place-based competition that Harvey (1989) predicted for entrepreneurial cities. Through the process of making the bid, Doctoroff became aware of large potential development parcels in the outer boroughs, including the Brooklyn waterfront, which informed the administration's economic development strategies. The Bloomberg administration took an aggressive stance toward rezoning and making available low-cost financing to developers in order to foster real estate development; as of 2009, fully one-fifth of the city had been rezoned (Buettner and Rivera 2009). Indeed, rezoning was one of the crucial strategies used to support the PlaNYC housing-related goals to increase the sheer amount of available housing stock, as well as to create affordable housing (City of New York 2007). Bloomberg's prodevelopment position included his endorsement of controversial and high-profile projects such as Hudson Yards on the west side of Manhattan, the new Yankee Stadium in the Bronx, Citi Field in Queens, and Atlantic Yards (now called Pacific Park) in downtown Brooklyn (Buettner and Rivera 2009; Brash 2011).

Some critics have argued that PlaNYC helped make Bloomberg's development agenda more palatable through efforts like green building incentives, brownfield remediation, and access to neighborhood open space (see table 2.1 for a list of the ten primary goals of the plan). Others disagreed; one nonprofit executive said that given that the mayor already had the "political juice to rezone half the city," PlaNYC appeared to be an earnest attempt at creating a more livable, environmentally sustainable city (respondent 34). Further, Bloomberg's policy agenda shifted from his first term to his later terms toward a more environmentally

TABLE 2.1 Ten goals of PlaNYC (2007 edition)

Land	• Create homes for almost a million more New Yorkers, while making housing more affordable and sustainable • Ensure that all New Yorkers live within a ten-minute walk of a park • Clean up all contaminated land in New York City
Water	• Open 90 percent of our waterways for recreation by reducing water pollution and preserving our natural areas • Develop critical backup systems for our aging water network to ensure long-term reliability
Transportation	• Improve travel times by adding transit capacity for millions more residents, visitors, and workers • Reach a full "state of good repair" on New York City's roads, subways, and rails for the first time in history
Energy	• Provide cleaner, more reliable power for every New Yorker by upgrading our energy infrastructure
Air Quality	• Achieve the cleanest air quality of any big city in America
Climate Change	• Reduce our global warming emissions by 30 percent

Source: City of New York 2007.

oriented disposition; he went from being the mayor that significantly cut the recycling program in 2002, to becoming the first to introduce recycling of all hard plastics and a citywide composting initiative in 2012.[4]

From the earliest days of his campaign, Bloomberg positioned himself as a mayor-CEO who would "run the city like a business" (Brash 2011). Yet because of the implementation of PlaNYC, commentators have claimed that his environmental work will be one of his biggest legacies, with Bloomberg known as a "green mayor" (Navarro 2009). These two claims may not be as contradictory as they initially appear. Even before it was identified as a sustainability initiative, PlaNYC had a focus on infrastructure development, land use planning, economic development, and support for the continued growth of the city in order to compete with other cities. In 2006, the city hired two consulting firms at a cost of more than $1.5 million to explore scenarios for how the city could accommodate a projected increase in population of one million residents by 2030. Following these reports, Doctoroff convened a series of high-level interagency meetings where city agency heads described different initiatives that would meet long-term goals, which led to the creation of the Strategic Land Use Plan in the summer of 2005.

The culture of work in this planning phase was characterized by practices brought over from the private sector. This included long workdays, multihour meetings with aggressive questioning of rationales, and demand for quantitative evidence. During the grueling twenty months in the lead-up to the launch of PlaNYC, public agency staff described being "put through the wringer" in four- to six-hour meetings with City Hall staff led by Doctoroff. Private consultants from the McKinsey Group were hired to conduct assessments of various sectors involved in PlaNYC—including transportation and energy—and to help with the development of the PlaNYC Sustainability Advisory Board. In addition, Rohit Aggarwala was selected to lead the PlaNYC effort. Aggarwala had previously worked at McKinsey and was selected for his quantitative reasoning, analytical style, and ability to lead this complex sustainability initiative. However, despite some critical scholars' claims to the contrary (Angotti 2010), PlaNYC is *not* just "a McKinsey plan." Numerous sources corroborated that McKinsey did not contribute to the Open Space chapter and that the goals articulated there were tied very closely to earlier proposals advanced by the Parks Department.

Under Bloomberg, Doctoroff, and Aggarwala, goal setting involved a careful political calculus regarding what was "actionable" rather than "aspirational." Amongst policymakers and staff that I interviewed, the juxtaposition of these two approaches was a consistent refrain. There was a shared sense that anyone could make a plan that would "sit on a shelf" and be ignored, but what made PlaNYC unique is that recommendations were selected and vetted for their "do-ability," with real funding allocations put toward initiatives. Thus, policymakers had a tendency to favor initiatives that could be monitored and evaluated quantitatively on an annual, quarterly, monthly, and even daily basis. Overall, the administration was committed to being "data driven" and focused on metrics.

On the rare occasions that PlaNYC went beyond the jurisdictional authorities of the mayor and the local public agencies, it met resistance that sometimes led to failure. The most visible example of this came through the proposed congestion pricing scheme for Lower Manhattan (inspired, in part, by that of London), which required state approval that it ultimately did not receive. Indeed, knowing that congestion pricing might potentially fail, the Bloomberg staff sought to balance that effort with some noncontroversial "easy wins"—such as the investment in citywide tree planting. One reporter claimed that Bloomberg had relatively more success with "smaller gestures" (Navarro 2009).

The administration began not only to develop the plan itself but also to embed it within city government for the long term. A *New York Times* article commended this institutionalization, noting that Bloomberg put PlaNYC "in the DNA of government" (Navarro 2009). In February 2006, a new agency was created: the Office of Long-Term Planning and Sustainability (OLTPS—now called the Mayor's Office of Sustainability), which provided strategic direction to the entire initiative and promoted interagency coordination in the long term. By June 2006, Aggarwala was appointed to head OLTPS and served until 2010, later replaced by David Bragdon. Bragdon characterized this office as different from a conventional planning agency in that it does not focus on spatial or land-use planning but rather on strategic planning. It operates "beyond the confines of functional agencies" and makes coordination amongst existing agencies its primary aim (Bragdon 2011). Of significance to the longevity of a plan that extended beyond Bloomberg's tenure, the city council passed local law 17 in 2008, which amended the New York City Charter and the administrative code. This law formally institutionalized OLTPS, the role of its director, and the presence of an advisory board. It also required annual reporting on a set of sustainability indicators (alongside the already existing Mayor's Management Report); updated population projections every four years; and updated sustainability goals and initiatives every four years.[5] While PlaNYC began as a mayoral initiative, it necessarily engaged with city council in a whole host of ways. It sought funding for new programs; it called for local laws and resolutions; and it recommended zoning changes. Thus, despite the strong Bloomberg imprimatur and heavy leadership from City Hall in the plan, the implementation of the various PlaNYC initiatives was a broader effort that engaged other facets of the municipal government.[6] Most central to the implementation were the various city agencies, to which I turn now.

City Agencies: Bureaucratic Authority and Influence

Building upon Lipsky's (1983) notion of "street level bureaucrats," Berg (2007, 244) highlights the crucial role of New York City's bureaucrats in shaping the implementation of policies and laws: "Bureaucrats whose function it is to implement and administer the law, and particularly those at the street level who interact with the public in the process of delivering a service, exercise considerable discretion in their implementation activities. Due to the choices that bureaucrats can make in the process of implementing the laws and programs of the city's political system, they can greatly influence the direction and shape as well as the success or failure of public policy." While the language of sustainability and the comprehensive scale of planning seen in PlaNYC may be new, municipal agencies in New York City have longstanding mandates to work on the issues covered in the plan: land use, water quality, air quality, housing provision, open space/parks, and recreation (City Charter 2009). Nearly every city agency could be found to have some link to this plan, as illustrated by the appendices to the plans that key each initiative to the relevant agencies (City of New York 2007, 146–55). The agency most directly engaged in forestry and gardening is the Department of Parks and Recreation (DPR).

DPR is one of the largest and most sophisticated urban natural resource management agencies in the country. The agency evolved out of the city's history of open space management, dating all the way back to the Dongan Charter of the City of New York in 1686 (NYC DPR 2013a). The 1989 charter revision mandated that DPR "manage and care for all parks" and "maintain the beauty and utility of all parks" (New York City Charter Revision Commission, Chapter 21, Section 533). In the absence of other legislation specific to park management, this gave the agency broad bureaucratic authority (Berg 2007, 245). As of fiscal year 2010, DPR had a staff of 7,242 full-time employees—including seasonal maintenance workers—and annual expenditures of $382.7 million (NYC MMR 2011). The broad physical scope of the agency's operations is described in the Mayor's Management Report:

> DPR maintains a municipal park system of more than 29,000 acres including nearly 1,800 parks, nearly 2,500 greenstreet sites, over 1,000 playgrounds, more than 800 athletic fields, more than 550 tennis courts, 54 outdoor swimming pools, 12 indoor swimming pools, 31 indoor recreational centers, 12 field houses, six community centers, more than 600 comfort stations, 14 miles of beaches, 13 golf courses, six ice rinks, five major stadia, 17 nature centers, 13 marinas and four zoos. The Department is also responsible for approximately 650,000 street trees and two million park trees, 23 historic house museums and more than 800 monuments, sculptures and historical markers. (NYC 2011c, 109)

The agency is divided into multiple administrative divisions, many of which are engaged in implementing different aspects of PlaNYC.

Urban forestry in New York City is primarily the responsibility of the DPR division Central Forestry, Horticulture, and Natural Resources, which was created in the late 2000s as a merger of Central Forestry and Horticulture (CFH) and the Natural Resources Group (NRG). CFH was directly responsible for convincing city leaders that a major citywide tree planting strategy could become one of the core components of PlaNYC. DPR manages the street tree planting citywide, through a system of central foresters, borough-based foresters, and city contractors who do the work of installing new trees.[7] NRG was founded in 1984 to manage New York City's "natural areas" or wild spaces—including "upland forests, maritime forests, meadows, fresh water and tidal marshes, lakes, ponds, and rivers" (NYC DPR 2010, 8).

Urban agriculture (specifically in the form of community gardening) is supported by the DPR division GreenThumb, although in a very different model from the way in which CFH is responsible for the urban forest. Prior to GreenThumb being managed via DPR, it was an office of the Department of General Services (now called the Department of City Administrative Services) that functioned primarily by offering short-term leases to community members as a means of dealing with the vacancy explosion in the 1970s. Created in 1978, GreenThumb offers basic material support, organizing, and technical assistance to more than five hundred gardens citywide, serves approximately twenty thousand gardeners, and is one of the largest community gardening programs in the country (Stone 2009; GreenThumb 2010). These gardens are community managed, but they are also supported by the City of New York through technical assistance and free materials—such as soil, compost, and plants (Lawson 2005; GreenThumb 2010). Although it is a program of DPR, GreenThumb continues to operate somewhat distinctly—particularly in its funding stream, which comes from federal Community Development Block Grants focused on low-income areas (Mees and Stone 2012). Indeed, there is no mention of community gardens in the City Charter's description of the authorities of DPR, although there is explicit mention of parks, squares, public places, and playgrounds (City of New York 2009).[8]

Along with DPR, several other city agencies are relevant to the implementation of urban forestry and urban agriculture. The Department of Transportation (DOT) has jurisdiction over the public right of way and is involved in the public plaza program, a PlaNYC initiative to create public gathering spaces in areas of underutilized roadway. The Department of City Planning, with heavy input from DPR, was involved in rewriting parking lot zoning regulations to require trees and in creating new development zoning regulations to require a dense spacing of new sidewalk trees (City of New York 2007). Though the New York City Housing Authority (NYCHA) was not prominent in PlaNYC, MillionTreesNYC leadership approached this agency early on, with interest in NYCHA grounds as potential planting sites for new trees. In terms of urban agriculture, the Sanitation Department historically provided leaf litter and compost to community garden sites, but this program was cut back under Bloomberg, and solid waste was not a key part of the initial version of PlaNYC (City of New York 2011a).

Numerous other agencies were involved in food policy efforts beyond the scope of urban agricultural production, instead focusing on food processing, distribution, sale, and consumption. The Department of Education (DOE) played an early key role in this effort through the school food procurement program, which was revised to incorporate more regional foods in public school menus. The Office of the Food Policy Coordinator was created by the mayor in 2007. Other local food policy documents called for the coordinator position to be expanded to a full New York City Food Policy Council, as discussed in chapter 6 (Stringer 2009, 2010; NYC Council 2010). One of the oft-cited challenges to developing and implementing a comprehensive food policy citywide is that so many of these functions are dispersed throughout numerous agencies—no one agency has ownership over this policy issue.

Civil Society: Advisors and Independent Actors

PlaNYC was also shaped by the precedents, projects, and partnerships initiated by many of the thousands of civil society groups in New York City. These include those groups formally involved in the Sustainability Advisory Board to PlaNYC (a mix of government and civil society representatives created in June 2006 at the same time as the handoff from Doctoroff to OLTPS).[9] One respondent lauded the advisory board for inviting potential critics into the inner circle of the process—including vocal members of the environmental justice community. Through working on the Sustainability Advisory Board, environmental advocates came to see Bloomberg as more than just "the guy who had canceled recycling," the respondent said (respondent 49). It is important to note, however, that neither the Sustainability Advisory Board nor consultation with select nonprofits played a role in setting PlaNYC's urban forestry–related goals. City Hall held targeted meetings with selected greening groups—including New York Restoration Project (NYRP), New Yorkers for Parks, and the Trust for Public Land—to vet emerging goals related to open space and greening. But because of DPR's clear leadership and expertise, the advisory board and these nonprofits played little role in crafting the forestry agenda. One decision-maker said, "It is fair to say that . . . the Parks chapter changed relatively little" (respondent 49).

Beyond the advisory board is a wider network of groups working independent of or alongside city agencies on urban environmental issues. For centuries, civic groups have played a crucial role in all aspects of the urban environment, including shaping the cityscape, influencing policy, and advocating for change (Svendsen 2010; Fisher et al. 2012; Connolly et al. 2013, 2014). Understanding the impact of New York City stewardship organizations on PlaNYC requires examining the role and attitude of key civic broker organizations (described in chapter 1) toward this effort. Although it was clearly a state-led plan, we can see the contours of how civil society groups shape—and are shaped by—the municipal initiatives that flow from the plan. Four main themes emerged from interviews with representatives of these civic brokers: the *process of PlaNYC's development;* the *governance of PlaNYC initiatives;* the respondent's *opinions of PlaNYC* overall, including critiques of what is missing from PlaNYC; and the *effects of PlaNYC on their organizations*. For the most part, these civil society groups supported these municipally led efforts and were eager to participate in new sustainability initiatives, while advocating for their expansion and improvement.

Civil society *involvement in the development of PlaNYC* ranged from complete disengagement all the way to formalized partnership in public-private collaborations. A number of broker groups said that they were not involved in any way. Other organizations were consulted, but their point of engagement ranged throughout the timeline of PlaNYC's development. One citywide greening nonprofit had an early consultation with Aggarwal and his team "before they knew what they were doing" (respondent 58). In contrast, a local conservation organization read the ten published goals of PlaNYC—which were released in December 2006—and approached the mayor's office with an interest in working together at that point. Another citywide organization focused on food and agriculture participated in the formal public engagement process that lasted from December 2006 to March 2007. One nonprofit focused on citizen engagement participated in lobbying and advocacy (along with 150 other organizations) in support of the congestion pricing initiative.

Governance of PlaNYC initiatives did, in some cases, take the form of public-private partnerships. The nonprofit Trust for Public Land formally partnered with the DOE on the Schoolyards to Playyards initiative in support of the PlaNYC goal that every New Yorker live within a ten-minute walk of a park, which included a commitment to convert 290 schoolyards into multifunctional community open spaces. Similarly, the MillionTreesNYC campaign was formally initiated via the plan, but the partner NYRP was not explicitly mentioned in the text of PlaNYC. A number of interviewees—speaking around 2011 or 2012—expressed concern about what would happen to PlaNYC after Bloomberg's final term. Some offered that further involving nonprofits in the governance and institutionalization of the plan would be one way to ensure longevity of the efforts.

Opinions of the plan were generally positive to mixed, with most critiques focusing on what was missing from the plan, rather than fundamentally questioning its intent. Those that held positive opinions noted that the plan would be part of Bloomberg's legacy forever, and they lauded the way in which the plan helped raise public awareness about sustainability. Some felt that the plan was neither positive nor negative and would not really affect the work that they pursued. Given that these respondents are all major nodes in the civic stewardship network, this ambivalence is meaningful. Critiques focused on the dilemmas that the dual commitments to housing development and open space creation might pose going forward. Many of the broker organizations felt that PlaNYC was a move in the right direction but that it missed certain key environmental issues.

Several noted that community gardens, school gardens, and food issues were entirely missing from the plan, although these areas had quite a lot of active interest from the public.[10] Other respondents noted that solid waste should have been examined; subsequently an entire chapter on solid was added in the April 2011 update. One organization involved in historic preservation noted that the plan focused more on new housing development than on building retrofits—with reuse being a much more environmentally friendly approach. Finally, that same respondent also noted that the issue of "green jobs"—meaning created paid work for environmental sustainability actions—was not adequately addressed in the plan. This reflected their position that the plan was not focused on social justice and equity but rather was "a plan for a million more people who work at Goldman Sachs" (respondent 57).

Interviewees identified the *effects of PlaNYC on their organizations*. Notably, a number of these broker groups had hoped that the plan would bring resources toward their organization, but this did not result. This contributed to the critique that the plan did not have sufficient funding for implementation. In contrast, ICLEI (2010b) praised PlaNYC for giving serious consideration to budgeting, funding, and financing the initiatives contained within it. This difference stems from one's vantage point: municipal budgets were certainly committed via PlaNYC; however, municipal funds were not generally used to fund private or nonprofit actors' participation. When public-private partnerships were formed as part of PlaNYC initiatives, they leveraged private resources rather than directing public funds toward the private or civic sectors. Beyond financial resources, multiple respondents noted that the existence of the plan shaped their own organizations' strategic planning. It helped focus their efforts and encouraged them to seek synergy with the mayor's agenda. Similarly, others noted that the plan encouraged them to create their own initiatives in response to it, such as neighborhood-level sustainability planning; grant programs related to storm water overflow; and holding public fora on historic preservation and climate change.

Whither the Public?

PlaNYC emerged out of several years of internal negotiation between City Hall and city agencies, leading to carefully wrought goals and initiatives. As such, the process was seen as secretive or closed, with public input coming only after key decisions were made (Barrett 2007; Mandelbaum 2007; Angotti 2010; Finn and McCormick 2011). One interviewee who was involved in the process from the early days of the strategic planning through the release of the official plan stated that city officials worked on goal setting for two years prior to the first public announcement. Another bureaucrat corroborated in stark language: "This was an *in house* enterprise. . . . It wasn't 'the grassroots.' The public was *not* involved. And then probably a year in, there was some time in 2006 that the public was finally allowed in" (respondent 41). Some insiders attributed the lack of broader public involvement to City Hall's goal of advancing the initiatives during Bloomberg's term. More time became available after Bloomberg pushed for overriding existing term limits to run for a third term, and the city council passed a law extending the number of terms that Bloomberg could hold. Advocates for good governance widely criticized this term extension. Yet many bureaucrats engaged in PlaNYC initiatives were grateful for the additional time to pursue their goals under a sympathetic administration.

After the internal agency work was complete, a six-month public outreach and engagement process was held, beginning with an announcement in December 2006 of the ten main goals of PlaNYC, followed by a three-month series of community listening sessions, in-person meetings, and digital fora from January to March 2007. These public meetings were criticized as tokenism: overly one-sided and lacking opportunities for meaningful involvement. One participant in the process went so far as to claim that all of the green elements of the plan were merely a "marketing overlay" for a growth plan for the city (respondent 52). Another interviewee described this process as an "unveiling, simultaneously to a lot of different constituencies: the press and the public at the same time and the special interest public" (respondent 41). A decision-maker involved in the formulation of the plan said that the administration was surprised and pleased by the enthusiastic reception that the plan received from the public. They had been prepared for much more opposition to the expenditure of public funds and instead received the opposite: "I remember at the announcement. . . . We invited all these environmental groups. And during the speech, people were crying. They were so happy. They were like, 'This has never happened before'—where they had submitted ideas and then so many resources were dedicated to it. And then the mayor was saying all the right things, like, 'These should be the long-term goals for the city . . .'" (respondent 26). In essence, the public engagement process helped get the word out about PlaNYC to those who were most inclined to care about sustainability, and perhaps to suggest minor shifts in how issues were presented, but not to substantively alter those goals. Given this context about the process behind PlaNYC, how, specifically, did an urban forestry agenda emerge and evolve?

PlaNYC's Urban Forestry Agenda

As cities around the globe mirror New York City's ambitious tree planting campaign, it is important to understand the origins of that effort. Locally, many public officials and nonprofit employees viewed MillionTreesNYC as one of the most successful of the PlaNYC efforts. At a May 2012 public event, Bloomberg called MillionTreesNYC "the flagship initiative of PlaNYC." The prominence of tree planting in New York City's sustainability plan invites investigation about the dynamics of agenda-setting. This section examines how the state and civil society negotiated the framing of issues and articulation of goals related to urban forestry in PlaNYC.

The Department of Parks and Recreation (DPR) Makes Its Case

DPR was successful in the PlaNYC goal-setting negotiation, as evidenced by the approximately $400 million capital commitments to that agency for MillionTreesNYC and the high degree of continuity between the initiatives that DPR proposed at the outset of the Strategic Land Use Plan and the official PlaNYC initiatives that were released in April 2007, which was not the case for all agencies. Skilled bureaucrats provided quantitative arguments to "lifer" DPR decision-makers in upper management who understood the political realities of working with City Hall.[11] Fiona Watt, chief of central forestry and horticulture (CFH), was ready with information her division had spent years collecting, organizing, and analyzing through data management systems, GIS and remote sensing studies, and a field-based tree census collected using staff and volunteers.

Both in interviews and in public remarks, policymakers described an "aha" moment that occurred between City Hall and DPR in crystallizing support for undertaking an ambitious, large-scale urban tree planting campaign. DPR officials argued for a massive increase in the budget and scope of the citywide street tree planting program. A bureaucrat described presenting scientific evidence of costs and benefits of urban tree planting to the mayor and his staff: "I think City Hall . . . [was] pretty persuaded by the science behind it . . . especially when a mayor comes in with a business background who likes everything to be quantified" (respondent 27). To make their case, DPR drew on local data coupled with the research of the U.S. Forest Service—in particular the STRATUM model. STRATUM offered a quantitative—and *monetized*—view of the urban forest, based on estimates of the value of various ecosystem services. Then-DPR commissioner Adrian Benepe was quoted in the *New York Times* as saying: "Trees are great for a variety of reasons, but how do you explain that to the Office of Management and Budget? . . . We plan on using these values as a baseline to say that this is what we have now, and argue for additional funds to plant more trees" (Randall 2007). DPR also knew exactly what it costs to plant a tree in New York City via its street tree contracting system. Overall, they found that trees in NYC represented a sound investment.

Decision-makers touted the quantified, multifunctional benefits of trees. Many of the economic benefits cascading from tree planting are associated with increases in local real estate value and commercial activity on tree-lined streets (Morales 1980; Anderson and Cordell 1988; McPherson and Simpson 2002; Wolf 2003, 2004, 2005; Donovan and Butry 2010). This appealed to the mayor—who viewed investments in green infrastructure and open space as part of a strategy to attract "global talent" to live and work in New York. Indeed, the STRATUM report to DPR is infused with the language of business sense and intercity competition:

> New York City's street trees are a valuable asset, providing approximately $100.2 million or $172 per tree ($15 per capita) in net annual benefits to the community. Over the years, the city has invested millions in its urban forest. Citizens are now receiving a return on that investment—trees are providing $5.60 in benefits for every $1 spent on tree planting and care. New York City's benefit-cost ratio of 5.60 exceeds all other cities studied to date, including Fort Collins, Colorado (2.18), Glendale, Arizona (2.41), and Charlotte, North Carolina (3.25). (Peper et al 2007, 2–3)

Moreover, knowing that trees make contributions to mitigating the urban heat island effect, improving air quality, and creating livable streets and vibrant commercial corridors (Nowak et al. 2010), policymakers could leverage different arguments with different constituencies. Some of the harder-to-quantify psychosocial benefits of trees were noted and used as arguments for the urban forest with the public, even if these were not the reasons that convinced City Hall. The notion that trees beautify neighborhoods, help reduce stress, and can promote more sociable, walkable street life were all harnessed as arguments—and later used in public relations campaigns (Hartig et al. 1991; Kweon et al. 1998; Kuo 2001; Kuo and Sullivan 2001; Taylor et al. 2001; Wells 2003; Orsega-Smith et al. 2004; Taylor and Kuo 2009). Furthermore, one decision-maker felt that trees were simply more accessible to and relatable for individual constituents than other sustainability interventions. He argued that many people could see, touch, and interact with trees in a way that was not as easy for, say, retrofitted green buildings.

Even debatable effects or highly complicated relationships were discussed as part of the suite of potential tree benefits. For example, a 2005 DPR program that predated the MillionTreesNYC effort was the Trees for Public Health (TPH) program, which targeted intensive tree plantings in areas with low street tree stocking levels and high rates of childhood asthma. TPH was developed, in part, on the premise that trees improve air quality and could potentially reduce asthma prevalence (Beckett et al. 1998; Nowak et al. 2006; Rosen and Greenfeld 2006;

Bealey et al. 2007; Lovasi et al. 2008). An introductory text to a section of PlaNYC about using "natural solutions to improve air quality" reads:

> Trees in particular are effective at cleansing the air. They do this by absorbing pollutants—sulfur dioxide, nitrogen dioxide, and carbon monoxide—through their leaves and intercepting airborne particulate matter on leaf surfaces. Every year, New York City trees remove an estimated 2,200 tons of criteria pollutants from the air. They also take in 42,300 tons of carbon each year. Indirectly, trees further reduce air pollution by shading buildings, thereby reducing the need for air conditioning during the peak electricity demand periods. In addition, shaded streets have lower temperatures in the summer, slowing the formation of ground-level ozone from NOX and VOCs. Trees also block wind in the winter, slightly reducing the need for heating. (City of New York 2007, 128)

The impact of urban tree planting on mitigating the urban heat island effect by reducing surface air temperature and lowering emissions from building energy use has a relatively clear scientific consensus (Nowak et al. 2010). The relationship between urban tree planting and air quality, however, is debated, as trees absorb some pollutants—such as ground-level ozone, particulate matter, and nitrogen and sulfur oxides—but can emit volatile organic compounds; and these effects can vary with the climate, built environment, forest composition, season, and time of day (Nowak et al. 2000, 2006, 2010; Pataki et al. 2011). Pataki et al. (2011, 32) note: "In general, the removal of atmospheric pollutants by vegetation is one of the most commonly cited urban ecosystem services, yet it is one of the least supported empirically." The urban forest-asthma linkage is even further complicated by numerous factors, such as income, age, indoor air quality, exposure to vehicular traffic, and access to health services (Jackson 2003; Krieger et al. 2005; Wu and Takaro 2007). Nonetheless, the PlaNYC goals and the MillionTreesNYC implementation retained the legacy of addressing and mitigating air pollution and—to a lesser extent—asthma.

Once City Hall was convinced of the desirability of a large-scale investment in tree planting, there still remained the work of enumerating a specific goal. A U.S. Forest Service study on New York City's urban tree canopy (UTC) found that the city had an average of 24 percent canopy coverage citywide and recommended that NYC should set a goal of 30 percent citywide canopy coverage by 2030 (Grove et al. 2005). However, PlaNYC decision-makers critiqued the percent canopy cover goal as presenting challenges in tracking, reporting, and communicating progress to the public, and DPR was not certain that a 30 percent goal was attainable.[12] A simple, numeric goal for number of trees planted could be measured for its progress each day, week, month, quarter, and year, and across all tree planting organizations, physical site types, and land jurisdictions. Setting such a numeric goal fit the requirement that the plan be "actionable" and "measurable," two of the core aims of PlaNYC decision-makers.

An Elite Nonprofit Sets Its Sights

Concurrently and completely outside of the Strategic Land Use Planning/PlaNYC process, a prominent nonprofit group in New York City became interested in developing a citywide tree planting effort. NYRP is a nonprofit greening organization that was founded in 1995 by celebrity actress and singer Bette Midler—and is one of the key civic broker organizations identified in chapter 1. At NYRP's 2006 spring picnic fundraiser, Midler announced that she wanted to plant one million trees in New York City. Numerous interviewees recounted this story as the abrupt starting point of NYRP's engagement with urban forestry at the citywide scale. Some speculated whether it was inspired by the Million Trees LA program, given Midler's part-time residence in Los Angeles. One respondent elaborated on the way in which the founder's changing interests shaped the programmatic focus of NYRP:

> NYRP, at the same time Million Trees was happening, was having an identity crisis because I think NYRP really didn't know what it wanted to be. It was doing lots of things. . . . This comes from the very leader of the organization, Bette, who—every day there's a priority and a new idea. . . . 'Cause when Bette picks up the phone and says, "Oh my gosh, I was just driving down 157th Street and there's plastic bags in the trees. Stop everything and get everybody up there." Now we're focused on plastic bags and then the next day it's something else. . . . NYRP is doing too many things instead of being really good at two things. (respondent 35)

The announcement surprised many of the long-time staff of NYRP and DPR, who did not view NYRP as having a forestry agenda or expertise. At the same time, Midler found support and enthusiasm for her idea in NYRP's newly hired executive director, Drew Becher, who saw the potential for the campaign to transform the organization.

Midler's celebrity offered her a platform for gaining an audience with officials at the levers of power. An insider noted, "Bette can definitely pick up the phone and talk to the mayor" (respondent 35). And vice versa, public officials were equally eager to partner with high-powered outsiders. The mayor established a track record of developing and funding innovative programs with prominent individuals, such as George Soros; and the DPR commissioner had a long tenure of nurturing public-private partnerships, such as the Central Parks Conservancy.

The sequence of what exactly transpired between Midler's pronouncement at the NYRP picnic and the city's formal announcement of NYRP as a partner in the MillionTreesNYC campaign remains unclear, but it centers on contact between Midler and City Hall. The story of a serendipitous collaboration between Midler and Bloomberg has become reified in its recounting. Midler tweeted on January 6, 2016, after the completion of the campaign, "So @MikeBloomberg & I turned to each other & said 'Let's plant 1mil trees in #NYC!' And so we did, and here's how: http://bit.ly/1SA1UmL." More specifically, one respondent claimed that Midler and Becher made a pitch for the million trees partnership with Deputy Mayor Patti Harris. Media accounts portray Harris as one of the key gatekeepers to Bloomberg. She is also noted for frequently offering private audiences to people in the arts world—in part because of her former role as executive director of the city's Art Commission (Steinhauer 2005; Freedlander 2010; City of New York 2012b). Another respondent noted Harris's centrality to the story: "She's a big piece of this. In city government, Patti Harris does a lot outside of the transparency of the city. She knows a lot of people. She can direct resources towards programs that aren't [in the] city budget" (respondent 13). After the declaration at the NYRP spring picnic, Bloomberg and his staff, Benepe, Midler, and Becher worked—initially via a closed door process—to craft a shared tree planting campaign.

The PlaNYC Forestry Goals Congeal

The goals and initiatives related to tree planting in PlaNYC reflected the multiple actors interested in tree planting and engaged in policymaking as well as the multiple benefits associated with tree planting that these actors used to justify investment.[13] The plan contained tree planting goals in the section devoted to open space but also in sections focusing on water quality, air quality, and climate change. See table 2.2 for a complete list of tree-related initiatives in PlaNYC.

In the open space chapter, there was a goal to "re-imagine the public realm" (City of New York 2007, 36). Specific initiatives included the effort to "green the cityscape" with the aim of raising the street tree stocking level to 100 percent by planting an estimated 23,000 street trees/year, reflecting the interests of the CFH division. While CFH was making its arguments for why the city should enhance investments in tree planting, other divisions (such as Community Outreach, Capital Projects, Natural Resources Group [NRG], Planning, and Parklands) were making their own claims. Indeed, the most heavily emphasized goal for the open space chapter was that every New Yorker should live

TABLE 2.2 Tree-related initiatives in PlaNYC (2007 edition)

CHAPTER	GOAL	INITIATIVE	TARGET ACTIONS
Open space	Re-imagine the public realm	Green the cityscape	Raise street tree stocking level to 100 percent by planting approximately 23,000 additional trees annually (p. 38) Expand the Greenstreets program by undertaking 40 new projects every planting season, bringing citywide total to more than 3,000 by 2030 (p. 38)
Water quality	Pursue proven solutions to prevent storm water from entering the system	Capture the benefits of our open space plan	Expand the Greenstreets program Increase the number of trees in the city by one million (p. 57)
	Expand, track, and analyze new Best Management Practices (BMPs) on a broad scale	Pilot promising BMPs	Plant trees with improved pit designs (p. 59)
	Expand, track, and analyze new Best Management Practices (BMPs) on a broad scale	Require greening of parking lots	Modify zoning resolution to include design guidelines for off-street parking lots that include trees and landscaping (p. 60)
Air quality	Pursue natural solutions to improve air quality	Capture the benefits of the open space plan	Support street tree planting by revising zoning code to require new construction and major redevelopment to plant one street tree/25 feet of street frontage Prioritize planting in neighborhoods with lowest stocking levels and highest air quality concerns (p. 128)
	Pursue natural solutions to improve air quality	Increase tree planting on lots	Reforest approximately 2,000 acres of parkland by 2017 at cost of $117 million (p. 128) Partner with stakeholders to help plant one million trees by 2017 (p. 129)

Source: City of New York 2007.

within a ten-minute walk of a park. This goal became the way that PlaNYC packaged up a wide range of initiatives developed with leadership from the then-head of the DPR Community Outreach division. This decision-maker saw the potential for trees in the streetscape to serve as connectors between larger park sites, with tree pits serving as "the smallest parks in the system" (respondent 41).

The chapter on air quality contained a goal of "[pursuing] natural solutions to improve air quality" via several initiatives, presented alongside a map of tree canopy ratios by neighborhood (City of New York 2007, 121, 128). In the first initiative of this goal, PlaNYC aimed to "capture the benefits of the open space plan." This initiative reiterated the commitment to 100 percent street tree stocking level but also (1) committed to revising the zoning code such that developers of new buildings or major renovations have to install one new street tree for every twenty-five feet of street frontage, and (2) provided an additional $17 million per year for 12,500 additional street tree plantings in TPH neighborhoods. A second initiative of this goal committed to reforesting two thousand acres of land citywide, building on the prior work of the NRG division (City of New York 2007, 128). Third, an initiative committed to "increase tree planting on lots" and specifically mentioned planting a million trees, without actually naming NYRP, as follows: "We will partner with stakeholders to help plant one million trees by 2017. The City will work with community, non-profit, and corporate partners on a 10-year goal to plant trees on private residential, institutional, and vacant land properties in order to achieve our goal to plant one million trees. The City and its partners will focus on areas whose natural environments have borne the brunt of past City policies, and neighborhoods with few green spaces" (City of New York 2007, 129).

One decision-maker at DPR went so far as to call the million trees reference "an accident," saying, "I don't think the City was really . . . prepared to commit to a million trees. . . . The goal to plant a lot of trees in New York City was in there [but] it wasn't defined as a million trees. . . . I think that sort of snuck through the editing process to tell you the truth" (respondent 47). This explanation of an "accidental goal" was *not* corroborated by other interviewees, who described the use of meticulous review and accounting for every goal by the mayor's staff as well as the use of a single writer in the later versions of the plan to help edit the document into one coherent voice. Nonetheless, there certainly was a sense that the million trees goal came in at a late hour and altered the course of goal setting that DPR and City Hall had been pursuing. While the million trees aim was threaded throughout several goals without naming NYRP explicitly, by contrast, the ten-minute walk-to-park goal was placed center stage and highlighted as one of ten key goals on the cover of the open space section.

Despite this, once the MillionTreesNYC campaign was announced and the plan was launched, there was no hairsplitting over origins. MillionTreesNYC was visibly branded as a mayoral initiative and one of 127 initiatives in PlaNYC. Some argue that the MillionTreesNYC campaign was used to help cement stronger public buy-in for the overall plan. In fact, the million trees goal was released to reporters one day in advance of the public release of PlaNYC on April 22, 2007 (Rivera 2007). The tree planting goal was viewed as a "feel good" issue, in sharp contrast to the political divisiveness of goals like the failed attempt at congestion pricing.

PlaNYC's (Lack of an) Urban Agriculture and Local Food Agenda

The urban forestry case clearly shows the implications of a high profile, top-down commitment by the mayor. In contrast, despite a long tradition of community gardening and an upswing in public interest in farming, PlaNYC's 2007 edition failed to mention urban agriculture or local food. Examined in the context of the history of food policy in the United States, this absence is unsurprising. At the federal level and in rural areas, food production and sale is regulated and incentivized as an agricultural commodity and market good; however, at the local level and in urban areas, food historically has not been a major aspect of the policy agenda (Pothukuchi and Kaufman 2000; Clancy 2004). Yet advocates claim that strengthening local and regional agriculture can enhance resilience, strengthen local economies, minimize environmental impact, and provide fresher, healthier, and better-tasting food. Speaking about this issue, one advocate said, "I would say food was the forgotten environmental issue. We spent . . . billions of dollars protecting our watershed. There are countless laws and regulations and programs to protect the air. . . . People need to breathe, they need to drink, and they need to eat. So, why shouldn't we have a rational food policy? . . . Why shouldn't we be thinking about the other thing we put in our bodies, which is food?" (respondent 34). Despite normative claims for why we *should* have a food policy, the question remains: why does New York City lack a comprehensive food plan? The failure of the City of New York to fully embrace food planning to the degree that certain other global cities have was seen by some city council members as directly due to the mayor's lack of strong commitment to this issue (Lander 2012; Levin 2012; Williams 2012).

New York City is not devoid of food-related policies, but rather PlaNYC lacked a single vision issued from the mayor's office. The Bloomberg administration did address important aspects of the food system through policy action—for example, creating a new position for a food policy coordinator (FPC). In addition, its

Department of Health and Mental Hygiene focused widely publicized policies on consumer behavior: food *consumption* and its links to obesity and diabetes. At the same time, food *production* and urban agriculture were not addressed. Indeed, the initial version of PlaNYC contained *no* mention of food systems, urban agriculture, or community gardening whatsoever.

Key public sector allies outside of City Hall worked with civic advocates to advance food policy and programmatic efforts, which, in turn, led to an increasing acceptance of food and agriculture as relevant issues. This includes the work of former city council speaker Christine Quinn and former Manhattan borough president Scott Stringer, as will be explored in chapter 6. Other city council members (including Leticia James, Melissa Mark-Viverito, Jumane Williams, Brad Lander, and Stephen Levin) also took actions like speaking at the Brooklyn Food Conference about the need for changes to the food system, lending support to local legislation on food issues, committing discretionary funds to projects in their districts, or supporting community gardens in preservation efforts. Respondents speculated on how broad-based changes in societal attitudes toward food—as well as generational changes in bureaucratic staff—had influenced policymaking by these local elected officials over time. An urban farmer said, "I think I would have even anticipated [food] plans like this a couple of years ago, given the consumer interest and the rise in Greenmarkets and the increased awareness of the foodshed in the tri-state area. It just took a couple of years, I guess, to get the young folks who were talking about this hired into the offices of elected officials. And then to get the elected officials to recognize that we'd get them public support and votes if they supported what the public thought" (respondent 44). This notion was corroborated by young staffers who felt they were given a great degree of latitude to help influence the agendas of the public officials for whom they worked.

A focus on local and regional food systems—including the stages of food processing, distribution, consumption, and postconsumption—gained more political traction in New York City than a focus on urban agriculture alone had. Many food policy efforts began with an emphasis on healthy food consumption but expanded to include broader issues as they evolved. In part, this was due to the physical limitations on space for agricultural production in New York City. It also reflected the size and robustness of the city's retail food and restaurant sectors; the amount of waste produced in the distribution, processing, and sale of food; and the multiple opportunities to effect changes through the system. Indeed, New York City's food distribution landscape is dauntingly complex, "comprising approximately 20,000 restaurants, 13,000 food retailers, 1,600 public schools, numerous hospitals, and other nonprofit service providers, as well as 90 farmers markets" (Barron et al. 2010, 10). This expanded view assembled unlikely allies under a common cause of improving the food system. Some food policy experts preferred to distance themselves from the perceived inefficiencies of urban agriculture and focused instead on other changes in the system—for example, developing regional food hubs, changing school food procurement rules, and strengthening upstate-downstate connections. Finally, the rising attention to food policy was also driven by interurban competition or policy transfer, with food plans emerging in Philadelphia, Seattle, Toronto, and London (for a survey of food planning in comprehensive and sustainability plans, see Hodgson 2012).

Bloomberg and the Food Policy Coordinator (FPC)

Bloomberg asserted leadership in the policy arena related to the public health dimensions of food. He created headline-grabbing policies that included a smoking ban in bars, restaurants, and later public parks; he instituted a trans-fat ban in 2008; and he required chain restaurants to post calorie counts on their menus. More controversially and with less success, he proposed to disallow the use of food stamps for purchasing soda, which was critiqued by advocates for the poor; he proposed a tax on sodas that was rejected by the New York State Legislature; and he proposed a ban on large-sized sugar-sweetened beverages, which was heavily opposed by the soda industry and delayed by the courts (Grynbaum 2012; Kliff 2012; Park 2012). These policies were divisive, with advocates celebrating them as proactive and cutting edge, and critics calling them examples of the "nanny state" overstepping its bounds (respondent 39).

Bloomberg also created the nation's first-ever food policy coordinator (FPC) in a major city, who oversaw the development and implementation of many of these policies. Irrespective of the actions the FPC undertook, just the *creation* of that position alone was seen by advocates as an important gesture toward institutionalization of food policy into municipal affairs. The office worked with the Department of Health and Mental Hygiene and the Department of City Planning to create the FRESH program to provide financial incentives to grocery stores to locate in neighborhoods with poor food access; the Green Carts program to create more mobile vendors of fresh fruit and vegetables; and the Health Bucks program to offer two-dollar vouchers for fruits and vegetables redeemable at farmers markets. An environmental nonprofit director commended the work of FPC and OLTPS in facilitating dialogue across silos, noting, "Before, in the food world there were the agriculture people, the health people, and the hunger people. And I think we're [seeing] an interesting intersection on those triangles in many ways" (respondent 34). Indeed, chapter 5 describes the various coalitions and discourses that were circulating around the multifaceted dimensions related to food, including sustainable farming ("agriculture people"), nutrition and diet-related diseases ("health people"), and food security ("hunger people"). Bringing these various framings and factions together through plans addressing the local/regional food system is described in chapter 6.

PlaNYC and Food

According to those involved in the early days of PlaNYC strategic planning, municipal leaders (including Bloomberg and his staff as well as Aggarwala and his staff) were not concerned with food—at least not relative to other issues that they were aiming to address. One decision-maker reflected on the challenges of assessing plan making and goal setting retrospectively: "To criticize [what is missing from the plan] is a lot of 20–20 hindsight. On food systems . . . food was not as big an issue then as it has become. . . . I think you have to look at it as a document of the time, in which case it was ahead of the curve by a little bit on climate change—certainly a lot on climate change adaptation. It was probably right with the curve on energy. . . . The biggest gap in PlaNYC was not relatively small issues like food, not . . . solid waste, but it really was that we took jobs for granted" (respondent 49). Another key PlaNYC policymaker noted that although food and agriculture issues were discussed, they were left on the cutting room floor because of a confluence of factors including lack of political champions and lack of sufficient information (respondent 26).

Food policy was seen at the time as largely outside the purview of city government because so much of the food system is beyond the geographic bounds of the city or controlled by the private sector. In this view, the system is too broad, complex, networked, and multiscalar for the municipal government to effectively address, and food itself is viewed as a commodity that the market can effectively distribute (see also Pothukuchi and Kaufman 2000; Cohen 2011). According to a researcher focused on urban agriculture and local food systems, OLTPS's geographically bounded jurisdiction limited its ability to enact transformative food policy changes. The agency had to be able to demonstrate quantitatively the impact of making changes to the food system—such as decreasing food miles—on the operation of the City of New York. He said, "Everyone's interested in the issue of regional food systems philosophically, ideologically, sympathetic, potentially. But what is the impact on New York City? . . . [OLTPS] saw it as out of their purview unless we're talking about decreased food miles within the city limit" (respondent 3). Making sense of the complexity of the food system and setting appropriate policy goals relating food systems and climate change was a computational challenge: "It's very difficult to designate the environmental benefits of food infrastructure changes. It's very easy to say that 'we have a building that's using x amount of kilowatts. And if we do y, the results are going to be z.'

It's much more difficult to say, 'If we change the food system in some way or other, our greenhouse gases are going to be reduced by x.' It's nearly impossible because it's just too complex" (respondent 34). Politicians face public pressure to demonstrate the impacts of policies, and this need is passed on to executive agencies that report directly to the mayor. Such agencies are simultaneously sensitive to the pressures of other elected officials, all of which leads to a frequent reliance on quantitative metrics to show achievements.

Overall, the food movement, visions, and plans were seen by municipal officials and staff as "aspirational," not "actionable." This was one of the more damning critiques from an administration focused on implementation, quantification, and policy that made "business sense." This notion is becoming reified within the policy discourse, as indicated by a 2012 report on food systems planning put forth by the American Planning Association: "When crafting plan goals and policies, balance and mirror aspirational goals with measurable objectives, indicators, and targets to enable effective plan monitoring and evaluation over time. Plan goals are often aspirational in nature. Therefore, in the absence of measurable objectives, indicators, and targets, local governments will struggle to evaluate progress in achieving the goals" (Hodgson 2012, 111). A food advocate noted the challenges of comprehensive food planning under the Bloomberg administration, critiquing the pursuit of "low-hanging fruit" policy goals: "Bloomberg's a businessman and we want metrics. We want to pick three to five things that we all believe we can get done. And that seems logical. But I think that when your system is so broken, we can't be looking for what's the big win or what's the low-hanging fruit . . . we have to be saying: 'we've got to do it all'" (respondent 14). For the authors of PlaNYC, the complex problem of the globalized food system and its associated challenges of nutrition, hunger, diet-related diseases, and equity did not mesh well with the plan as a local-scale municipal infrastructure and land use strategy.

PlaNYC and Agriculture

Although open space was included in the plan, the chapter focused on traditional recreational parks, the public right of way, and "natural areas" but did not include any focus on agricultural spaces. There were no goals, policies, targets, nor even any discussion on urban agriculture. Indeed, the words *community garden*, *farm*, and *agriculture* were not mentioned in the plan, and there was just one mention of *roof gardens* in the context of a green roof tax incentive. One official involved in plan making said, simply, "It did not come up" (respondent 47). This absence was broadly noted by agriculture advocates; many said they were not involved in any way in the development of PlaNYC. Some felt that gardening and farming could be seen as "quaint" or antiquated by decision-makers and not associated with the new green city of the twenty-first century that New York sought to be (respondent 31).

Throughout the crafting of PlaNYC, a variety of narratives emerged about the spatial politics behind urban agriculture: sites either were too small and diffuse to matter citywide or, conversely, were in competition with housing development. Some respondents noted that urban farming was seen as ad hoc, inherently experimental, and difficult to scale. The small size of existing community garden sites contributed to their being overlooked by DPR leaders in the early PlaNYC goal setting. According to a bureaucrat, "[Community gardens received] virtually no attention at all. And there's a very simple reason. This exercise was an ambitious, game changing opportunity. . . . Now all of the community gardens in the city put together—it's maybe forty acres of land. We have a nearly 30,000 acre park system. . . . In the context of PlaNYC and talking about what will have an impact on this huge shortfall of land available for park purposes, [community gardens] wouldn't have moved the dial at all" (respondent 41). Bloomberg's pro-development position meant that the administration would never consider offering developable land away for gardens or farms. Identifying potential sites for development was as much a part of PlaNYC as any of the open space, air, water, or climate change initiatives. The vacant developable parcels that *were* available in 2007 were thoroughly reviewed as sites for development of affordable and market rate housing, as part of PlaNYC's housing goals. According to a public official, "I think one thing the Bloomberg administration did amazingly well was to really document where land was. When I came to [city government], we didn't even know what we owned, where our sites were. And we really did a massive job of inventorying land—there's just not that much unprogrammed land" (respondent 50). Officials were cognizant of the historic perceived tradeoff between housing and gardens that culminated in the garden auction and crisis of the late 1990s, and they sought to avoid such a conflict going forward (see chapter 5).[14] One key official involved in PlaNYC goal setting said that community gardens "came up in a couple of conversations," but officials perceived a "zero sum game between housing and gardens" (respondent 49). It is clear that policymaking with respect to gardens was not occurring in a vacuum and must be placed in the context of the prior twenty years of development, community organizing, and management of these sites.

Given the administration's commitment to quantification and metrics, urban agriculture was critiqued—both by public officials and by advocates seeking to influence decision-makers—for a "lack of data." While quantifying the amount of vacant, sunny, potentially arable land in the city is a challenge, projecting the potential agricultural output from those sites is a greater challenge, and calculating the holistic costs and benefits of those sites is an even greater endeavor still. Many of the demonstrated benefits of urban agriculture come not from the amount of produce grown, but from the provision of a unique form of open space with opportunities for community engagement, education, employment, and neighborhood improvement. Advocates noted that these social benefits remain difficult to quantify. A policymaker commented that the mayor was "less interested in the softer elements" of urban agriculture, particularly its linkages to youth employment and community empowerment (respondent 39).

In the absence of a strong, personal commitment to urban agriculture on the part of the mayor, it is imaginable that an organized constituency could have pressured and created the needed political will, if they had enough size, power, and influence. However, several interviewees felt that the current movement around urban agriculture and local food systems had not yet coalesced in 2006–2007. Others believed that the constituencies were in existence but were not recognized by decision-makers, or lacked the knowledge of or access to the channels through which to make their priorities known. One official critiqued these advocates for their lack of savvy about the policymaking process:

> I think that [food and agriculture advocates] are probably the most curmudgeonly folks in the whole world, because there's this sense of, "Well we've already figured it out. You just need to get on board." . . . So, the foodie world is not as politically sophisticated as it could be. . . . Like there's this big beef about . . . setting aside empty public lots for community gardens. The city's not just going to say, "Yeah, you can have all these community gardens." You know, some residents don't have homes. . . . How do you [show] that you *understand* what the city has to think about . . . so that you can tailor your argument in a way that makes sense to them so that they'll hear you? I don't think that the foodies have done that as effectively as they could have. (respondent 52)

The approach that advocates used to pressure the administration and the way in which policy agendas shifted over time, including in the PlaNYC2.0 update, will be discussed in chapter 6. During its formative years, however, PlaNYC made no mention of the constituencies, issue areas, or policies related to urban agriculture and local food systems.

Whose voice is heard and which issues decision-makers view as salient and politically feasible all inform which items make it onto the sustainability agenda. Forestry had vocal internal advocates at DPR, coupled with vocal external advocates in civil society who were poised to push forth a public-private partnership. Urban agriculture, as part of a local and regional food system, was a complex issue without a single powerful agency to advocate for its inclusion. City leaders more readily took up policies related to healthy food consumption rather than the politically less palatable or less feasible efforts around urban agriculture. Comparing what gets included in and excluded from the sustainability plan helps render visible the multidimensional operation of power in creating urban nature. From this starting point in 2007, to what extent and in what ways did the sustainability plan get implemented? Where do we find progress or change in how the sustainability agenda treated forestry and agriculture? How did agendas shift over time through political acts, discursive framings, and material practices?

3

CITY OF FORESTS

Planting One Million Trees

The MillionTreesNYC campaign was a citywide effort to plant and maintain one million new trees from 2007 to 2015, a program that has become a globally recognized model for urban forestry initiatives.[1] In under a decade, this campaign succeeded in its goal of greening a quintessentially gray, concrete city. In order to understand *how* one million new trees were planted in New York City, we must unpack the actors and material practices involved in the campaign. Indeed, New York City's urban forest was institutionally and physically transformed through new governance arrangements, organizational changes, counting routines, and material practices.

This chapter's narrative starts after the sustainability goals of PlaNYC were set, examining the implementation of the campaign. A public-private partnership was created to run the campaign, an example of hybrid governance, with blending of roles and responsibilities in pursuit of a shared goal. The prominence and scale of the initiative led to organizational changes in the two core partners as they both garnered and expended massive amounts of resources in the planting of a million trees. The conduct of urban forestry in New York City was also radically altered by the campaign. New procedures were developed to alter the material practice of urban forestry on streets, in parks, in "natural areas," on public housing grounds, and on private land—the entire matrix of green spaces was considered as plantable area in pursuit of the million trees goal. Across all landscape types, every effort was made to maximize the tree canopy through successful planting and survival of trees, despite challenging urban site conditions.

While tree survival was a priority, increasing biodiversity was another crucial aim—particularly in the restoration and afforestation practices that emphasized growing healthy, native, multistoried forests—as was equity, by addressing spatial unevenness in street tree distribution.[2]

Hybrid Governance of the Complex Urban Environment: Public-Private Partnership

The MillionTreesNYC campaign was launched as a joint effort on October 9, 2007 (see figure 3.1). The creation of a public-private partnership between the City of New York and NYRP leveraged the assets of both sides to support the campaign. The strategy of using public-private partnerships was one "hallmark of the Bloomberg administration" (respondent 24). In an ideal view, the expertise and economies of scale of the large municipal agency would be balanced by the nimble innovation of the professionalized nonprofit organization. First, DPR brought its long track record of large-scale street tree and park tree planting as well as forest restoration work in "natural areas." This included systems

FIGURE 3.1 Leaders of the MillionTreesNYC campaign at the 2007 public launch. From left to right: Drew Becher, former Executive Director, NYRP; Adrian Benepe, former Commissioner, DPR; Michael Bloomberg, former Mayor of the City of New York; Bette Midler, founder of NYRP; Patricia Harris, former First Deputy Mayor of the City of New York. Courtesy Malcolm Pinckney, DPR.

for sourcing, purchasing, transporting, installing, and maintaining trees in collaboration with a network of private contractors. Second, NYRP brought access to corporate and individual funders that were not as easily accessed by DPR. Third, NYRP was recognized for its savvy with marketing, public relations, and outreach campaigns—and for its ability to rapidly innovate in ways that a city agency could not. Finally, the ability to plant on both public and private lands was required in order to meet the ambitious tree planting goal.

DPR and NYRP established formal and informal linkages between two institutionally and culturally distinct organizations. They developed a memorandum of understanding (MOU) that delineated the specific terms of the partnership. Signed in September 2008, it articulated roles and responsibilities related to "tree planting and care, education and outreach, marketing and public relations, urban forestry research and program evaluation" (MTNYC MOU 2008, 2). The MOU specified that DPR would plant street trees and on DPR land, while NYRP would plant on private land, housing campuses (NYCHA grounds), schoolyards and playgrounds, and other "publicly accessible private lands" (MTNYC MOU 2009, 4). Although not written into the text of the MOU, partners agreed on an explicit planting ratio, according to respondents, with DPR planting 60 percent of trees and NYRP planting the other 40 percent. However, over time, the ratio shifted to 70 percent DPR and 30 percent NYRP.[3] The MOU also established a $35 million fundraising goal for the campaign for NYRP (MTNYC MOU 2008, 7). While the MOU was a signed, written agreement, it was not legally binding. Nonetheless, it is clear that the shared governance of the campaign was based on carefully negotiated, formalized agreements and not just informal networks of reciprocity or trust described in the idealized version of network governance (see Davies 2011).

The parties routinized their contact through a number of means: constant email contact, monthly meetings about planting decisions, biweekly meetings between the directors of MillionTreesNYC at DPR and NYRP, and biweekly joint meetings called the MillionTreesNYC Taskforce. Some respondents saw the taskforce as an important means of communication, but others argued that these meetings consisted of reporting back tree planting metrics rather than making joint decisions. Despite this contact, a respondent leveled the critique that MillionTreesNYC was a strained—even divided—partnership: "The first and most important challenge was, despite my cries and pleas, that we had two teams running one program. And that is the crux of every problem that that initiative faced. [DPR was] doing things in their office and had their own set of priorities and were having their own set of conversations with funders and partners and other government agencies. And then you had NYRP doing the same thing. . . . I mean, if you have a football team and you have two sets of coaches giving directions . . . it's a joke" (respondent 35).

Despite their deeply entwined relations, the two sides differed in *mandate* and *objective*. Several interviewees felt that DPR had a higher level of accountability to citizens because it is part of the municipal government. This sense of the public trust was crucial to DPR's ability to successfully meet the targets of the campaign. NYRP, in contrast, was motivated by a "moral obligation"—particularly the passions and commitments of its founder (respondent 28). These fundamental differences necessitated mutual learning and acceptance. Government agents learned that civil society had great ability to generate outside resources and to engage with the public and the media in unique ways without the hurdles required by municipal oversight and regulation. And vice versa, the nonprofit sector, which can sometimes critique government as slow or inefficient, found they had much to learn about how to roll out a complete, citywide tree planting program.

The divide between the two sides was also a result of differences in capacity and scale of the organizations. One interviewee discussed the different sizes of the partners:

> I mean, Million Trees is this two-headed monster sometimes in that we really work at creating a partnership so that there is a cohesive program being delivered. But a non-profit organization of eighty people . . . it's significantly smaller than the thousands of employees that the Parks Department has, right? You have, on one side, a government agency with thousands of employees, huge infrastructure, sort of like military, para-military management. And then you have a non-profit organization that's much smaller in size that relies on outside entities to fund it. . . . (respondent 15)

The learning curve for NYRP was so large that it presented challenges in the face of a demanding, metrics-driven, mayoral priority. Even when DPR provided NYRP with technical assistance, resources, and training in urban forestry practices, the differences in capacity of these groups as *tree planting entities* were stark. Although NYRP was not selected as a partner because of its forestry expertise, it fell short of DPR's expectations for its ability to successfully plant large quantities of trees citywide on a short timeline. One leader critiqued NYRP's role saying, "[NYRP was] much more strong on the marketing and education side. The branding side of things. I'm not sure they've actually done very much in terms of planting trees or even planting trees right . . ." (respondent 27). Framed another way, however, these differences can be seen as an opportunity: "So you had people who really knew how to plant and grow trees in the city combined with people who really knew how to market and create a narrative about it. . . . And both groups had 'skin in the game.' They needed to prove to the mayor's office that they could produce. . . . And I think that was a really exciting moment for both these groups" (respondent 28). Overall, DPR and NYRP differed in size, mission, institutional structure, governance, and capacity. Their complementary skills and approaches were leveraged in a formalized partnership that jointly ran the campaign, but not entirely without tension.

Organizational Transformation: Expanded Funding and Staffing Changes

After the launch of the campaign, both sides of the partnership responded to increased budgets and staff size associated with MillionTreesNYC. One of the key factors shaping DPR's intensity of commitment to the campaign was the capital funding allocated to the agency through PlaNYC.[4] PlaNYC allocated capital funds for fiscal years 2008–2017 and operating (expense funds) for fiscal year 2008. Specifically, it committed $226 million in capital for the street tree planting goal and $150 million for the forest restoration goal (NYC DPR 2011).[5] While many of the stated PlaNYC initiatives did *not* come with such massive budget increases attached, DPR received hundreds of millions of dollars to fund tree planting and park enhancements. Indeed, one respondent noted that this capital increase represented the largest increase to the DPR budget since the "WPA and Robert Moses era" of park development (respondent 15). An internal memo echoed this sentiment, saying, "PlaNYC represents the most significant change in municipal urban greening since the Parks Department first funded citywide curbside tree planting under Robert Moses in 1934" (Watt 2007). For a list of DPR funding from PlaNYC, see table 3.1.

PlaNYC's capital commitments necessitated rapid progress on tree planting, as there are challenges associated with using capital dollars committed over several fiscal years. One interviewee said that any funds in a capital budget allocated for future years but not yet spent were "funny money" that could be lost at any time (respondent 41). Capital monies can be rescinded, and even projects in contract or with work started can be halted based on decisions made by the city Office of Management and Budget (OMB) and City Hall. In fact, the DPR NRG division, which was responsible for forest restoration, initially failed to appropriately spend its capital resources and ended up losing $11 million out of their budget back to the city OMB general fund. Other than this setback, DPR worked aggressively to allocate and spend all the committed capital dollars. This was done with the knowledge that DPR's budget is malleable and often receives cuts in the back-and-forth between the mayor and the City

TABLE 3.1 DPR's PlaNYC Funding, with MillionTreesNYC-related capital funding in bold

CAPITAL BUDGET	
8 regional parks	$386 million
290 open schoolyards	$96 million
36 field lighting sites	$42 million
25 synthetic turf fields	$22 million
800 greenstreets	**$15 million**
2,000 acres of new forest	**$150 million**
220,000 street trees	**$226 million**
Total capital budget for PlaNYC projects	$906 million
ANNUAL EXPENSE BUDGET	
7-year pruning cycle	$2.7 million
Stump removal	$2.0 million
Maintenance staff (227)	$10.4 million
Total annual expense budget increase for PlaNYC projects	$15.1 million

Source: NYC DPR. 2011. "MillionTreesNYC-PlaNYC." Presentation. May 16, 2011.

Council, both of which are hesitant to make cuts to schools, police, or fire department services.

The increases to the budgets also translated to an ability to hire more staff. DPR, like many municipal agencies, is governed by a complex set of rules related to hiring city employees. Although capital funds are devoted to long-term capital improvements to the city, up to 10 percent of capital dollars can be used to fund staff to oversee design, contracts, and installation. Expense budgets are also used to hire staff (but these were not increased as dramatically by PlaNYC, see table 3.1). Hiring municipal employees is a slow process, as is the reorganization that occurred to bring NRG into the CFH division and to fill behind positions after people left the agency. Particularly notable was the departure of Fiona Watt from her leadership role; that vacancy had ripple effects throughout the agency for several planting seasons. Despite these complexities, DPR did have large staff increases—particularly in the street tree planting division, which went from a staff of less than ten to approximately twenty-five. In addition, the MillionTreesNYC office within DPR did not exist prior to PlaNYC—this division consisted of a full-time director, deputy director, and project coordinator, as well as a volunteer coordinator hired through AmeriCorps who later became a seasonal DPR employee.

On the NYRP side, this nonprofit found itself having to learn very rapidly how to serve as a citywide urban forestry organization—a role it had not previously filled. The organization's initial focus was on improving the quality of parks in Northern Manhattan (such as Fort Tryon Park, Fort Washington Park, and Highbridge Park), which was the direct result of Midler's viewing these parks on her commute into and out of New York City. Later, in the mid-1990s during the community garden crisis under Mayor Giuliani, NYRP bought more than fifty gardens that were going to go to auction as housing development sites, thereby expanding the organization's focus to include gardens.[6] A staff member noted this most recent organizational shift, saying, "Now, truth be told, we had not planted that many trees before. So we were, in some ways, a very odd fit to the initiative. . . . It was a big, big leap for this organization" (respondent 10). The engagement in tree planting was a bold step into new terrain, a jump in scale to citywide work, and also an opportunity to transform the organization. Indeed, the campaign brought the organization national attention and name recognition.

NYRP underwent massive structural and tactical changes as a result of growth from MillionTreesNYC. From 2007 to 2010, the organization went from a budget of $6 million to one of $13 million, which is an astronomical leap for the local nonprofit environmental sector. One respondent estimated that NYRP added at least ten full-time staff focused solely on MillionTreesNYC. As with any large organizational change, the transition was not entirely smooth. Some interviewees argued that the leadership did a poor job of informing both upward (to the board) and downward (to the staff) of the full implications of this campaign for the organization. The increased attention from funders and the public on MillionTreesNYC led to a reorganization of existing programs within NYRP and the scaling back of prior programs focused on things like boat building and community gardens. Overall, MillionTreesNYC "dramatically moved [NYRP] from being perceived as just a cleaning and greening organization to an organization that had become a little more sophisticated in terms of understanding policy and the impact that has on decision-making" (respondent 62). With that maturation came a change in how NYRP related to other organizations, including DPR:

> Well, I think it's safe to say that earlier on in our work we've had adversarial relationships with organizations that we also now have good relationships with, Parks Department being an example. When you're a young, arrogant organization and you go in and you do work without telling them . . . or when you publicly say in the paper, "Well, Parks Department can't take care of it so we came in and saved the day," it's like saying to your neighbor, "Your yard looks awful so I am going to tell the whole neighborhood that we went in and cleaned it

> up for you. Do you like us now?" "Well, no, not so much. Don't call us out." (respondent 62)

As this example makes clear, through this joint campaign the organization shifted tactics from an advocacy role focused sometimes on confrontation, to a shared resource management role focused on collaboration.

Physical and institutional terrain that previously had been almost solely the responsibility of the municipal government came to be shared with a new suite of players, including philanthropies, corporations, and local nonprofits. One of the main roles of NYRP (and the Mayor's Fund) in the campaign was soliciting outside funds. Examined positively, private resources helped supplement an ambitious public campaign.[7] Examined critically, the philanthropic community provided a funding source that was less transparent and was outside of public oversight and approval processes. The rationale for having both NYRP and the Mayor's Fund involved was to be able to court the broadest suite of potential donors. Crucial to jumpstarting the campaign was a $10 million funding commitment made in the spring of 2008—first $5 million committed by Bloomberg on behalf of Bloomberg Philanthropies, which was then matched by $5 million from David Rockefeller and The Rockefeller Brothers Foundation. These donations reveal the role of the billionaire mayor as private actor, who could bring external, *personal* resources and networks to help guarantee the success of his signature initiative. Those $10 million were given via the Mayor's Fund, which then distributed the funds primarily to NYRP for its direct-planting component of the campaign. NYRP worked to develop relationships with corporate entities to support the campaign and eventually secured the three lead sponsors: Toyota, BNP Paribas, and Home Depot. In addition to making major donations, corporations were solicited by both NYRP and DPR to engage in corporate volunteer tree planting days. In sum, the campaign catalyzed both a reconfiguration of relationships involved in the governance of the urban forest as well as organizational growth on both public and private sides.

Counting Practices: Managing, Monitoring, and Messaging about the Quantified Forest

The move from the U.S. Forest Service's percentage canopy cover goal to the PlaNYC goal of planting one million trees was a significant shift brought about by multiple forces. The 30 percent canopy goal came from scientific recommendations in order to mitigate ozone-related air quality in the city (Luley and Bond 2002; Grove et al. 2007). However, staff in the Mayor's Office were skeptical that a 30 percent goal was achievable in the highly developed environment of

New York City and wanted to set a different mark. Having run various analyses, they found 27–28 percent to be achievable, but not 30 percent; and they did not want to set a goal that they could not meet. In essence, pragmatism altered a goal that originated in environmental science. Moreover, City Hall believed that a simple numeric goal would be more legible to the public and trackable by managers. One interviewee said, "No one knows what the hell tree canopy is" (respondent 35); another called a million trees a "sexy number" (respondent 48); and another acknowledged, "It's also a nice round number just in terms of marketing and branding" (respondent 1). Responding to all of these issues, DPR staff made "back of the envelope" calculations to show that one million new trees were roughly equivalent to the 6 percent canopy increase that the city sought to achieve. With this stroke of calculation, the two different types of goals were merged into one single goal: planting a million trees.

The MillionTreesNYC campaign had a consistent focus on counting, tracking, and reporting the number of trees planted. This is not surprising, given that the numeric goal was so tied to the identity of the effort and given the metrics-driven nature of both the Bloomberg administration and philanthropic donors. The pressure to count increased because the campaign was time-delimited as a ten-year effort from 2007 to 2017. Furthermore, both DPR and NYRP staff were acutely aware of an effective end date of 2014—the change in mayoral administration. One public sector employee said, "It feels like . . . there's a ticker on Bloomberg's desk. Next to revenues and cost are the million trees and he's watching each ticker turn. If that thing isn't turning or on schedule, someone's going to hear about it" (respondent 13). This pressure reflects the reality of trying to make progress on a long-term sustainability plan with goals and impacts that stretch to 2030—in the context of a mayoral term that lasts four years. The fact that Bloomberg secured a third term—though mayors were previously limited to two terms—was politically controversial and critiqued by many as antidemocratic but was appreciated by leaders of the MillionTreesNYC campaign for giving the effort more time to achieve its goals under a supportive administration.

This focus on counting disciplined both internal management and external public relations of the campaign. Internal operations focused on this accounting daily, weekly, monthly, quarterly, annually, and over the lifetime of the campaign. Mechanisms used for this tracking included daily phone calls from NRG staff to the MillionTreesNYC office, a weekly review of the DPR horticulture report by the MillionTreesNYC office, and the reporting of supervisors back to DPR on the number of trees planted by their contractors. In addition, biweekly meetings between DPR and NYRP focused, in part, on tracking the number and location of trees planted. One of the program staff maintained a master spreadsheet tracking all of the trees counted toward the total tree planting figure. This figure included "direct plantings" by the campaign partners; "tree giveaways" to homeowners and the public; and "influence plantings." Of these figures, the "influence planting" calculation was the most ambiguous. It was calculated by counting the number of trees sold at area retailers, such as Home Depot and Lowes, and discounting that figure by 75 percent. Maintaining a precise count of trees planted obscured these estimation practices.

Public events and outreach strategies reinforced the focus on counting. For example, at the launch of the campaign, the first symbolic tree (planted with the help of Bloomberg, Midler, and *Sesame Street*'s Big Bird) was marked with an oversized tag counting it as "one in a million." A digital ticker counted the number of trees planted on the home page of the campaign's website, and people who planted trees on their own property were encouraged to register those trees online and contribute to the total count. The planting of the 500,000th tree on October 18, 2011, was celebrated with great fanfare, as was the planting of the 800,000th tree on November 14, 2013, toward the end of the Bloomberg administration (City of New York Press Release 2011b).

This focus on counting said nothing about the survival or condition of the trees, nor about the process by which they were planted. Trees that were planted by residents on private property were occasionally self-reported online, but their condition was not verified in the field. For trees planted by NYRP on publicly accessible private lands in the first few planting seasons of the campaign, before better operational routines had been developed, there was an overemphasis on speed. Moreover, though the mission of MillionTreesNYC was "to plant *and care for* one million new trees," focusing on the number of trees planted makes no claims about the net number of trees that survive. NRG "natural area" plantings—with their numerous small trees—make a significant contribution to the total planting figure. Yet forest growth and competition among plants naturally involves some tree mortality, which is both expected and planned for in the management of these sites. Many of these sites were challenging areas to restore or afforest, with young native trees competing to survive with invasive, weedy species. This led one natural resource manager to dispute the numeric goal:

> Obviously it looks really good to have a million trees get planted for the mayor. But from our budget point of view . . . I don't think that it was the smartest decision because we can't really *maintain* most of these trees very well. . . . I mean, our mortality rates according to our statistics aren't terrible but a lot of sites don't do as well as they should. . . . The smarter thing from an ecological point of view would have been to plant *not* a million trees and then just maintain the ones we did a little bit better. . . . Set a lower goal but then just get more *quality* out of that. (respondent 48)

Thus, the metric for NRG sites might more appropriately have been "acres of forest treated/restored," but the numeric goal of trees planted carried over across all landscapes: street trees, park trees, forested areas, and private homes—even when the logic begins to break down on forested sites. Still, every attempt was made to maximize survival rates on reforestation plots, and 468 permanent research plots were created to monitor survival over time. Thus far, monitoring and evaluation has found that 88 percent of trees survived the first season, and of those that survived, 90 percent survived the second seasons (Simmons 2010, unpublished dataset, cited in Campbell and Monaco 2014). In addition, the monitoring of long-term outcomes of New York City forest sites that were restored in the 1990s has demonstrated improvements in species diversity, greater forest structure complexity, and evidence of the regeneration and retention of native tree species, as compared to unrestored sites (Simmons et al. 2016). As the campaign matured, it revealed a progression from a focus primarily on planting to an even larger emphasis on stewardship and maintenance. Respondents noted this in the case of NYRP as well as DPR, particularly through the creation of the Stewardship Corps (StewCorps) and later the TreeLC program, which are discussed in chapter 4.

Material Practices and Physical Transformation of the City: Street Trees, Natural Areas, and Private Land

The urban forest is a complex assemblage of the built form of the city, the trees themselves, the labor of people, and institutional structures. Understanding the way in which trees are planted and maintained in New York City requires examining differences in these processes by property jurisdiction (public, private) and site type (street trees, park trees, forested lands). The challenges related to site selection and installation reflect the reality of altering nature in a densely built urban environment. A forester quipped, "Urban forestry is not forestry" in acknowledgement of the challenges of infrastructural conflicts, human interactions, building shadows, and traffic (respondent 17). Urban forestry is a professionalized practice with its own conventions, rules of thumb, and best practices for managing the urban forest. Although DPR is often the lead entity involved in New York City, groups from all three sectors—public, private, and civic—are enmeshed in the process.[8] In order to understand the changes wrought via MillionTreesNYC, this section examines how the urban forest is materially constructed across *street trees in the public right of way (PROW)*, *"natural areas" in forested parks,* and *landscaped private lands,* bearing in mind the actions of human managers as well as the needs and capabilities of trees themselves in these various sites.

Street Trees in the PROW

According to the initial PlaNYC goal, DPR aimed to plant 220,000 new street trees in the PROW over the course of the MillionTreesNYC campaign, which was a major jump from their prior pace of planting. At the outset of the campaign, DPR held a two-day conference and invited dozens of experts in the worlds of arboriculture, horticulture, and open space management to advise the agency on best practices for such a massive urban tree planting effort. Information gathered in this meeting, as well as a series of internal memos from the then-head of CFH Fiona Watt, was used to inform changes to DPR's street tree management practices. Table 3.2 summarizes the changes that occurred to urban forestry in the PROW as a result of the MillionTreesNYC campaign. In addition to changes to

TABLE 3.2 Changes to planting and maintenance practices of street trees in the PROW

	PRE-MILLIONTREESNYC	MILLIONTREESNYC
Procurement	Landscape planting company under contract to DPR purchases trees from nursery.	DPR contracts directly with nursery, which grows trees to their specifications. Significant expansion of species palette, more quality control of stock.
Managing tree requests	Request-based system leads to uneven forest cover across neighborhoods.	Environmental justice approach: block planting in Trees for Public Health neighborhoods plus request-based system means planting trees in areas with lowest stocking level first. Policy change that private residents cannot refuse tree siting in the PROW in front of their homes.
Assessing planting sites	Average pit size 25 square feet.	Create the largest pit possible to increase soil volume and rooting space. Recommended pit size 50 square feet, where possible. Large expansion to DPR forestry staff involved in site assessment.

(Continued)

TABLE 3.2 Continued

	PRE-MILLIONTREESNYC	MILLIONTREESNYC
Planting trees	Planting on streets conducted by city contractors. Individuals, builders, businesses, and business improvement districts (BIDs) can plant by permit.	Zoning changed to require developers to plant a tree every 25 ft of street frontage of new or renovated buildings. DPR contractor planting continues.
Stewardship and maintenance of trees	DPR block pruning program provides street tree maintenance since 1998. Street tree stewardship occurs mostly through independent action or programs of TreesNY, Brooklyn Botanic Garden, or other nonprofits as well as through New York Tree Trust.	New citywide citizen stewardship programs developed: StewardshipCorps, TreeLC, mini-grants, adopt-a-tree. Eventually led to the creation of Director of Stewardship position within DPR. Young Street Tree pruning program created.
Removing trees and tree valuation	City removes dead trees, hazard trees, and downed limbs. Public can mulch Christmas trees. Removal of street trees allowed under DPR permit that requires replacement of basal area.	Unrelated increase in removals due to extreme storm events during MillionTreesNYC. Local law 3 in 2010 codifies NYC Tree Valuation Method for replacement of trees under DPR jurisdiction.

procurement practices, policies, and programs described in detail below, the MillionTreesNYC era also brought about an important change in local legislation. Local law 3, passed in 2010, is a tree restitution law that requires the application of specific replacement criteria when DPR issues a removal permit. So when a tree is removed under DPR's jurisdiction, it must be replaced by a Tree Valuation Formula that is set forth in DPR rules. While not applicable to private lands, this law acknowledges the ecosystem services of all city trees and institutionalizes a mechanism for their replacement.[9]

Since 2008, street trees have been contract-grown specifically for the City of New York in nurseries throughout the mid-Atlantic region, giving significantly more quality control over planting stock to the municipality (Watt internal memo 2007). While trees were once purchased on a season-by-season basis by tree planting contractors, since MillionTreesNYC, nurseries have entered directly into a business relationship with the City of New York itself, growing trees exactly to the city's specifications. The list of acceptable tree species and cultivars that can be planted on New York City streets was significantly expanded post-PlaNYC, in an effort to enhance the biodiversity of the urban forest (see http://www.nycgovparks.org/trees/street-tree-planting/species-list).[10] Species selection reflects ecological, aesthetic, and practical considerations—hardiness and transportability are major factors in determining which species survive well in the urban forest. Once trees are sufficiently matured to a minimum 2.5-inch caliper size (the diameter of the tree 12 inches above the ground), the trees are excavated, their roots are "balled and burlapped" in a contained root ball, and they are loaded onto flatbed tractor trailers for delivery to New York City (see figures 3.2 and 3.3). At every stage, DPR oversight is built into the process, from selecting and tagging trees at the nursery to being on-site during deliveries. One respondent talked anthropomorphically about the care given to the trees: "We deliver the trees . . . literally they're like children. It's being taken to a nursery school. While they're at school the other people are in charge of them. And they're ensuring them. And they have to feed them and bathe them and everything that's required. It's like handing off a treasure" (respondent 27).

Prior to PlaNYC, DPR used a request-based system to select tree planting sites. Given a limited amount of resources to plant trees, tree requests were collected primarily via the city's 311 call center.[11] This centralized 311 system was set up by

FIGURE 3.2 Street trees being delivered by flatbed truck. Photo by author.

FIGURE 3.3 "Balled and burlapped" street trees for planting. Photo by author.

Bloomberg as part of his reorganization of city operations to focus on providing better services to residents as "customers" (see also Brash 2011, 75). The rationale behind this planting approach was the assumption that a resident requesting a tree would have a greater likelihood of engaging in stewardship practices, or at the very least would not be hostile to a tree (Watt internal memo 2007). One unforeseen result, however, was the production of an uneven urban forest. This unevenness was rendered visible in new ways through the data collected via DPR's 2005 Street Trees Census and through the use of GIS mapping of 311 tree request calls and tree planting locations (Rae et al. 2010). Certain neighborhoods (often more affluent and educated neighborhoods, such as Park Slope in Brooklyn) showed a much higher volume of requests—and, therefore, more trees planted—than areas that were less affluent or educated (such as East New York in Brooklyn).

Because of this challenge and prior to the conception of PlaNYC, DPR created the Trees for Public Health (TPH) program that focused on block planting—planting entire city blocks with street trees—in six neighborhoods with the lowest street tree stocking levels and the highest incidences of childhood asthma and respiratory diseases.[12] However, since the time that TPH was created, the scientific thinking about the relationship between trees, air quality, and asthma shifted (as described in chapter 2). Nonetheless, DPR managers continued to emphasize the low street tree stocking levels as the key neighborhood selection criterion for block planting, while also noting that the locations of high childhood asthma prevalence in New York City are highly correlated with low income, high minority, and generally underserved areas. Thus, TPH block planting advanced an equity agenda. NYRP echoed this thinking and focused their private planting efforts for MillionTreesNYC in the TPH neighborhoods first. This aligned with NYRP's mission to focus on "underserved neighborhoods, communities of need" (respondent 11). A natural resource manager explained that PlaNYC funding created the opportunity to continue the block planting and equity approach that had been started under TPH. She described 311 tree requests as a "bottom-up approach to tree planting" and the block planting as a "top-down" approach (respondent 4). Block planting concentrated PlaNYC resources in the hopes of making a more immediate, transformative, and visual impact on neighborhoods, rather than solely filling in missing tree pits here and there.

With the funding from PlaNYC, DPR's mandate was to plant *every available street tree planting opportunity.* Foresters described selecting the "right tree for the right site" as the greatest challenge in urban forestry (respondents 17, 18). It requires a mix of art and science, knowing which trees will perform well in specific local conditions, as well as an aesthetic judgment about what mix of trees will be visually appealing. The DPR training manual was revised in 2008, but foresters also reported a high degree of learning on the job through fieldwork to assess the immediate microenvironment of potential tree planting locations.[13] Foresters in the city do not actually plant trees; they select sites for tree pits. The minimum recommended pit size was increased post-PlaNYC both because of the demonstrated impact of pit size on tree survival and because of the idea that tree pits could come to be seen as "tree beds"—as sites for stewardship and engagement or as "the smallest parks in the system."[14]

PlaNYC also led to the creation of a new citywide policy aimed at supporting equity in the distribution of the urban forest in the PROW. As of the spring 2008 planting season, residents did not have the right to refuse having a tree sited in front of their home because that denial was seen as taking away from a collective good. This means that street trees are viewed as an infrastructure no different from a stop sign, which cannot be refused. DPR made the case for this policy shift that was then instituted at the mayoral level, which ensured that the policy was upheld and enforced. Foresters are trained to interact with the public by explaining trees as green infrastructure and to head off potential challenges before they escalate. Out on the streets, residents who choose to talk to foresters tend to have strong opinions—polarized between highly supportive and highly opposed to trees. Supportive residents are often those who requested trees in the first place. Those opposed are often concerned with the potential for sidewalk damage, trip hazards, requirements to keep sidewalks clear of debris including falling leaves, and even possible damage to homes, roofs, basements, and sewers. On rare occasions, resident opposition to trees can intensify to the point of physical resistance, vandalism, and the removal of trees (Susman 2009). Part of this resistance comes from the ambiguity over who "owns" the PROW—with some seeing street trees as the responsibility of the government to maintain and others seeing the space in front of their homes as their property. Although residents do not technically have property rights to the sidewalk, they do have a responsibility to maintain it free of debris and snow. This site literally becomes a "grey zone," where the public and private intermix (Rae et al. 2010). See figure 3.4 for an image of a street tree with signs of stewardship and claims of tree ownership on handwritten signs.

New York City street trees are planted by landscape contractors working for DPR.[15] Trees are guaranteed for the first two years of their lives on the street under the terms of the contract, meaning that if a tree dies within that time, the contractor is obligated to replace it. This extra focus on watering and care for young trees is an effort to ensure survival, since after the first two years the tree is more established and thought to be better able to withstand the variations in temperature and rainfall that inevitably occur. After the two-year contract guarantee, street trees are left to survive with baseline maintenance provided by DPR. There are programs for addressing storm damage, sidewalk damage, and block pruning, but there is a backlog of requests for sidewalk repair, and maintenance budgets are often the first to be cut. As of fall 2011, the pruning cycle lengthened to a fifteen- to twenty-year cycle, such that fewer street trees were pruned each year (Foderero 2011). However, DPR continues to refine its maintenance program, including through the creation of a new Young Street Tree Pruning program in 2012. Since 2013, managers reported that the pruning cycle has averaged seven years; and as of 2014, the "Trees and Sidewalks" budget for sidewalk repairs was doubled.

For all of PlaNYC's capital commitments to tree planting, managers, critics, advocates, and allies alike felt that it needed a greater commitment of funds for maintenance and stewardship, which was made more acute following cuts to the maintenance budget in 2008. One municipal employee said, "Nobody wants to be the mayor of a city with a million dead trees" (respondent 1). In addition to the DPR block pruning program that had been in place since 1998, MillionTreesNYC leaders explored different models to cultivate residents' volunteer stewardship of street trees. They developed online Adopt-a-Tree websites and tree-care pledges and gave away free tools for gardening in tree "beds." As of

FIGURE 3.4 A stewarded young street tree with signs claiming it as private property in Carroll Gardens, Brooklyn. Photo by author.

November 2012, however, just 5,506 trees were adopted citywide, and approximately 1,500 stewardship actions were reported online (MillionTreesNYC 2012). A campaign leader noted that many more trees were adopted than were reported online, but there was no clear mechanism for tracking that activity. The StewCorps program, described in chapter 4, developed stewardship training and certification in partnership with several nonprofits throughout the city and later evolved into the TreeLC program. Developing grassroots stewardship of trees citywide was, according to numerous respondents, one of the greatest challenges to the campaign.

Starting in 2014, public attention focused more intensely on the survival and health of trees planted through the MillionTreesNYC program. A reporter at the *New York Times* wrote an article focusing on research about the campaign, noting, "Already, researchers at the [NYC Urban Field Station] have determined that street trees planted during the campaign had a survival rate of 93 percent after the first two years, compared with the 89 percent before the program" (Foderero 2014a). The public radio station WNYC used DPR data accessed through a Freedom of Information Act (FOIA) request to create an interactive map of street tree mortality (http://project.wnyc.org/dead-trees/). The authors raised questions about the success of the MillionTreesNYC campaign, citing a 12.1 percent mortality rate as contrasted to DPR's stated 6.7 percent mortality rate (Ye and Schuerman 2014).[16] Later that summer, Comptroller Scott Stringer conducted an audit of DPR's pruning contracts. Stringer's office drew attention to poor tracking and contract oversight, particularly in Manhattan and Staten Island—leading to trees not being pruned that needed it and other trees receiving pruning that did not require the work (Stringer 2014). This report subsequently led to media coverage on the radio and in newspapers, both local and national. Finally, the research community has drawn repeated attention to issues of tree mortality through studies in New York City, comparative studies, and the development of national standards through the Urban Tree Growth and Longevity working group (see, e.g., Lu et al. 2011; Roman 2014). All of the procurement and planting standards and practices have been established to maximize tree survival, but there will always be challenges experienced with growing and maintaining trees surrounded by cement on the sidewalks of New York City.

"Natural Area" Forested Parks

Approximately two-thirds of the first 500,000 trees planted in the MillionTreesNYC campaign were sited in so-called natural areas through afforestation and reforestation practices. These natural areas are managed by NRG—a division of DPR—but can occasionally include land under the jurisdiction of other agencies, such as the Department of Transportation (DOT). The sites are more rugged or "naturalistic" and are not programmed for active recreation. While many of these sites were held for years by DPR, they were more or less unmaintained until 1984 when then-DPR commissioner Henry Stern created NRG. Mimicking the language of New York State forest preserves, Stern created a city "Forever Wild" designation in order to draw attention to "natural" sites in the DPR portfolio. These fifty-one sites include woodlands, meadows, marshes, and wetlands that are often adjacent to infrastructure (such as highway interchanges, parkways, airports, and large facilities) or on sloped or rocky land not suited to other recreational uses. They are often larger in acreage than typical recreational parks and comprise more than 8,700 acres citywide (City of New York DPR 2012; City of New York 2011a). The soils on this site often include construction debris and other fills, which affects the chemistry and biology of the soil and, therefore, the ecosystem that it can support. Finally, as with many urbanized and fragmented sites, they are heavily colonized by invasive species—such as mugwort, porcelainberry, ailanthus, and multiflora rose (see also Johnson 2012). Because of their size, condition, and historic lack of funding and public attention, a manager described NRG sites as a "bottomless pit of need" (respondent 36). Figure 3.5 shows a large tree at a forest restoration site covered with invasive vines.

FIGURE 3.5 Tree and invasive vines in forest at Cunningham Park in Queens. Photo by author.

The original PlaNYC goal for "natural areas" called for reforesting two thousand acres citywide. In a nonurbanized context, two thousand acres is a relatively small area of forest. However, in the context of the highly developed city, identifying two thousand viable acres to reforest was an ambitious and technically challenging goal, or as one respondent called it, "a fool's errand"—since this amount of unprogrammed, open space acreage did not exist in New York City (respondent 65). DPR immediately scaled up its existing invasive removal and native tree planting practices. In order to meet targets for the number of trees planted each season, the agency began identifying known sites that were large and ready for planting. Concurrently, DPR took two actions to harness the knowledge, skills, and resources of a broader set of actors.

First, DPR created a $937,000 contract with the environmental design firm EDAW (now a part of the firm AECOM) to help guide the forest restoration efforts. Administered as part of a $9.9 million capital contract for reforestation, the design contract was originally supposed to last just six months, although the timeline was extended several times. EDAW's primary tasks were to identify potential planting sites citywide to contribute to the two thousand–acre goal (using GIS and field-collected data); develop an approach for prioritization among those sites; design and implement three pilot reforestation sites; and develop a "cookbook" of design and planting best practices to guide that work. Respondents reported difficulties on both sides in terms of communication, leadership, unrealistic timelines, and confusion over intent of deliverables—particularly the idea of a cookbook set of approaches when every site faces unique challenges, opportunities, and management needs. Although the contract began at the very beginning of the MillionTreesNYC campaign, the final forest restoration guidelines book was completed by DPR in 2014, reflecting the scope and magnitude of a document that went far beyond a cookbook (City of New York DPR 2014).

Second, NRG and EDAW both brought in outside expertise to help inform the reforestation effort and to facilitate the use of sites in research. In September 2008, EDAW convened a symposium of natural resource managers and scientists focused on reforestation. At that meeting teams began to articulate research questions and opportunities for "designed experiments," whereby the planting effort would also be used to generate new knowledge about urban ecology. Rather than a single unified citywide approach, at least three distinct research efforts emerged to study planting and management practices via slightly different planting strategies, plot designs, and sampling techniques—two of which were included in the EDAW work.

Some managers expressed concern that NRG sites were viewed by decision-makers as just green spaces on the map that could receive hundreds of thousands of trees. From the outset, many advisors and land managers took issue with a numeric tree planting goal and argued that emphasis should be placed on creating healthy, native, multistory forests. Over time, however, the campaign messaging around forest restoration shifted from an emphasis on the two thousand–acre goal to an emphasis on the million trees goal. One interviewee offered the critique of the million trees goal: "It blew my mind over and over again that you can be this silly, aiming for a number that has nothing to do with feasibility" (respondent 65). Another respondent said, "Well that's the other thing that changed is we pretty much know we can't make two thousand acres. At least not with the money we have. And we could easily [plant] 480,000 [trees] and not get to two thousand acres. And that's something that we just don't even mention any more. . . . There was at some point an acknowledgement that we couldn't do it, at least from NRG's perspective. And Parks knows it, but I don't think that's at all publically acknowledged" (respondent 4). As the campaign crystallized, tracking trees planted became *the* metric by which progress was measured. While DPR's current tracking of forest plantings indicates mortality rates of 10–12 percent, some forest ecologists with whom I spoke projected that over time, hundreds of thousands of trees planted in forest sites will die through natural forest competition. With young trees planted densely, it is to be expected that some of these saplings will die off, while others will quickly grow to close the canopy.[17]

Despite the limitations of the numeric planting goal, NRG staff were able to use PlaNYC resources to help achieve their ecological goals. They built upon restoration plans that were established in the 1990s but never fully funded, and they used the funding to plant interstitial spaces adjacent to parkways and infrastructure. The management approach to these sites involves clearing invasive vegetation and planting young, native trees in close proximity in order to close the tree canopy and prevent the return of invasive species.[18] One major limitation of PlaNYC tree planting funding in the context of "natural areas" is that it was set aside only to be used for *tree* procurement and planting. So although managers were seeking to restore forests with canopy trees, understory shrubs, and an herbaceous layer, the funding could only be used for trees. Managers worked creatively to find other sources of funding that could be used to help develop the multistoried forest. One solution they developed was to contract with the DPR-managed Greenbelt Native Plant Center (GNPC) to grow shrubs and herbs for forest restoration sites. In addition to growing these materials directly, GNPC also provided seed stock and tree liners to private nurseries that were in procurement contracts with DPR to ensure that these plant materials were native to New York City. See table 3.3 for a summary of changes to management practices in park "natural areas."

TABLE 3.3 Changes to planting and maintenance practices in natural area forested parks

	PRE-MILLIONTREESNYC	MILLIONTREESNYC
Reforestation sites and scope	Reforestation occurs through grant funding and bond monies in specific, limited areas across the city from the 1990s.	Citywide attempt to identify 2,000 acres for reforestation. While this acreage was not identified, approximately 2/3 of the first 500,000 trees planted were located in natural areas.
Procurement	DPR or contractor purchase trees and plant material with DPR or grant funding.	PlaNYC funding can only be used for tree planting—DPR staff must use other funds to plant understory and herbaceous layer. Seed source for tree stock must be within 200–400 miles of New York City in order to create native forests. DPR's Greenbelt Native Plant Center used as seed source for tree contracts and for shrubs and herbaceous materials.
Experimentation	Practices of trial and error leading to established routines and best practices.	Best practices continue, but selected sites are planted explicitly with experimental design (of vegetation palette and management treatments, with control plots) and monitored over time by researchers and managers.
Planting	Done through a mix of in-house staff and private contractors.	Volunteer planting program created, along with staff and contractor role. Grows to where sites can be planted with 20,000 trees in a morning using volunteers.
Maintenance	Done through a mix of in-house staff and private contractors.	Natural Area Volunteers program created to engage volunteers in maintenance, along with staff and contractor role; program eventually rolled into citywide Stewardship Programs.

Publicly Accessible Private Lands

NYRP had responsibility for planting on anything other than public property. With some echoes of the process that NRG faced in trying to identify acreage citywide in which to do reforestation and afforestation work, NYRP set about trying to identify large landholders who would be amenable to tree planting. Early on, they identified university campuses—particularly the City University of New York (CUNY) system—and public housing grounds (or NYCHA "campuses") as potential candidates.

The relationship with NYCHA was complicated from the start. Although NYCHA houses approximately half a million people in 334 developments citywide (NYCHA 2012), the agency was included neither in PlaNYC's strategic planning processes nor in the goals that emerged. As one interviewee noted, "NYCHA missed the boat on PlaNYC" because of its unique administrative structure—unlike other city agencies, NYCHA is a New York State public-benefit corporation with a seven-person board appointed by the mayor, federal funding, and relative autonomy from routine city operations (respondent 7). Once NYRP became aware of the 2,600 acres of open space citywide in NYCHA's lawns and plazas, they viewed NYCHA as a key planting site. Mirroring the approach used in pitching the campaign to the mayor's office, NYRP went straight to upper NYCHA management and sought a blanket license agreement to plant trees on NYCHA's grounds.

With a license agreement secured, NYRP was able to plant on NYCHA grounds using its field staff, Americorps members, and corporate volunteers. The NYRP model of bringing in large volunteer groups for single-day planting events was not modified for planting on NYCHA grounds, despite their having a history of isolation, marginalization, and complex—often racialized—politics. Concurrently and separately from PlaNYC, NYCHA was engaging in its own green planning following the 2007 appointment of Margarita Lopez as NYCHA Environmental Commissioner. This resulted in the cultivation of Resident Green Committees—voluntary tenant groups in developments that would engage in recycling, energy conservation, education, and stewardship efforts (NYCHA 2011). NYRP adjusted its approach and did outreach via these committees, seeking to engage them in planting and stewardship efforts (with varying degrees of success). Despite this, Lopez made public remarks critiquing the approach to tree planting used on NYCHA grounds and argued that NYCHA residents should be given green jobs and paid to take care of these trees. After a few summers of planting, the relationship between NYCHA and NYRP leadership (Lopez and Becher, respectively) began to strain. One interviewee called it "toxic" (respondent 62). Nonetheless, as of fall 2011, NYRP had planted nearly twelve thousand trees on NYCHA grounds (NYCHA 2011).

NYRP also planted trees on other publicly accessible private lands, such as the large lawns of Co-Op City in the Bronx. Co-Op City is a privately managed housing development with more than fifteen thousand residential units in thirty-five high rises and another seven townhouse buildings spread out across three hundred acres, including a sixty-acre greenway through its center (Whitsett 2006). Learning from the process of working with NYCHA, NYRP later sought to negotiate agreements with both senior management as well as grounds staff, or "the suits and the boots," in order to ensure that all divisions were onboard with planting and maintenance (respondent 13). On October 1, 2011, a large-scale planting day was held at Co-Op City where approximately 300 volunteers planted 250 trees. These planting arrangements continued to be negotiated on a partner-by-partner basis, reflecting the complexity and patchiness of land ownership and jurisdiction in the city.

Private Lands—Tree Giveaways and "Influence Plantings"

The final ways in which the MillionTreesNYC campaign worked to grow the urban forest was by offering free trees to individual residents and by tracking plantings done by other outside entities (besides DPR and NYRP) who were "influenced" by the campaign. First, NYRP organized tree giveaway events at parks and gardens, often in partnership with a local community group that could assist in promoting the event throughout the neighborhood (see figure 3.6).[19] At these events, residents were limited to one tree per household and were given instructions on tree care. In the spring of 2011 alone, 3,500 trees were given away. These events were seen as a way to build awareness about trees, facilitate planting on private land, and develop support for the campaign amongst residents who received trees. However, there was no guarantee that these trees were planted, planted correctly, or survived. A DPR forester noted that sometimes these trees end up being planted "guerilla style" in empty tree pits on the street, where often they do not survive. Because these giveaway trees were not grown or installed to New York City's street tree specifications, they can be removed by the city when they are detected by foresters.

"Influence plantings" were perhaps the greatest area of ambiguity in terms of accounting for the impact of the campaign, as these are plantings that are *neither* funded *nor* carried out by DPR or NYRP. As part of this attempt to influence other entities and engage them in the campaign, NYRP offered outreach sessions to architects, planners, landscape architects, and developers. In addition, information was collected from developers and private landowners (such as IKEA in Brooklyn, which went through a park development effort along with the construction of its building). Trees planted by these entities were also counted.

FIGURE 3.6 Tree giveaway organized by NYRP and Gowanus Canal Conservancy. Photo by author.

Finally, as mentioned, MillionTreesNYC received sales information from area tree retailers, such as Home Depot and Lowes, and estimated that 25 percent of trees sold in a season ended up being planted in New York City.

The MillionTreesNYC campaign was created to handle the scale and complexity of New York City in expanding the urban forest of a global city. The hybrid governance arrangement supplemented government resources with those of the private sector and made sites on both public and private land accessible to planting. Counting practices rendered the campaign legible to decision-makers and the public, even if these acts of quantification masked some of the ambiguities of tree planting and survival. New operational routines and partnerships aimed to maximize successful expansion of the urban forest on streets, in parks, in forests, and on institutional and residential yards. But physical planting and maintenance were just the start of MillionTreesNYC; the campaign hoped to engage residents, organizations, and businesses as full supporters of the urban forest.

4

BEYOND PLANTING

Creating an Urban Forestry Movement

Striving to be more than just a tree planting campaign, MillionTreesNYC sought to create an urban forestry movement through messaging, partnerships, and engagement opportunities. Movement-building entailed cultivating support across municipal leadership, among civic groups, and with the public in order to fund the program, plant and steward trees, and sustain the newly expanded urban forest into the future. Examining the narrative framing of MillionTreesNYC and the storylines deployed shows how leaders positioned the campaign to policymakers and the public to help garner top-down support and bottom-up participation. The already-existing network of stewardship groups in New York City responded—and in some cases reorganized—because of the creation of such a prominent citywide campaign. DPR and NYRP worked with a broad network of partners via the campaign's advisory board to create programs with impact on education, stewardship, green jobs, and research—all of which expanded the reach of the campaign beyond planting alone. The prior chapter focused on *how* one million trees were planted; here we attend to *why* one million trees were planted by exploring the discourse of the campaign—as well as why the campaign resonated with such a large set of actors citywide. Finally, challenges and opportunities arose over the course of this multiyear effort. Substantial changes stemmed from the 2008 global economic recession, the maturation of the campaign, and leadership turnover at DPR, NYRP, and the mayor's office.

Frames, Storylines, and Messages: Multiple Benefits of the Urban Forest

From the beginning, PlaNYC treated trees as a form of green infrastructure designed to help efficiently address the needs of a growing city.[1] Entrepreneurial decision-makers saw investing in trees as a strategy to help enhance the livability of the city and compete with other cities for businesses and residents. When asked about the rationale for planting a million trees in New York City, leaders of the campaign said that trees can help mitigate some of the negative side effects of a growing city and can make the city a more livable, sustainable, and competitive place. They remained "on message" with the claims put forth in the PlaNYC document about trees and aligned themselves with urban sustainability as an overarching concept. A brief excerpt from the campaign's website reflects this storyline about the green, growing city and the role of trees in it:

> **Why Plant a Million Trees?**
> New York City is *growing!* You can see it—and *feel* it—in every neighborhood in every borough. It's exciting, and it's what makes New York the greatest city in the world. But, like in any thriving metropolis, it's important to make sure the Big Apple and its residents—meaning *you!*—are healthy and happy while adjusting to the growth and the many changes it will bring with it.
>
> Planting trees is one of the most *beneficial* [links to a page about environmental, economic, and health/lifestyle benefits of the urban forest] and cost-effective ways to help ease these growing pains. Trees help clean our air, and reduce the pollutants that trigger asthma attacks and exacerbate other respiratory diseases. They cool our streets, sidewalks, and homes on hot summer days. Trees increase property value, and encourage neighborhood revitalization. And trees make our City an even more beautiful and comfortable place to live, work, and visit (MillionTreesNYC 2010, emphasis original).

The campaign started from an assertion that one million more trees were good for the city, citing research about the multiple, quantified benefits of the urban forest. This storyline fits, in part, with a neoliberal understanding of the urban forest as an amenity that creates value for landowners and that can be harnessed into giving the city a "green" image (see also Heynen and Perkins 2007; Perkins 2009).

Nonetheless, it is important not to discount the themes of ecosystem health, human health, and quality of life—including more intangible benefits like beauty and neighborhood aesthetics—that the campaign embraced and deployed. One manager said that the rationale behind the campaign was twofold: enhancing the growing city but also embracing "a growing recognition of the value of nature and natural systems" (respondent 36). While City Hall may have been convinced by economic arguments, the campaign had a flexible discourse that was broad enough to incorporate diverse actors with distinct motivations for engaging with the campaign—whether they be members of the public, particular constituent groups, or funders. One nonprofit employee noted the strategic nature of this claims-making when speaking to funders: "It's saving energy and shading children on playgrounds when you're talking to funder X. . . . It's connectivity of habitat and bio-diversity for funder Y. And then for funder Z, it's connecting people to nature" (respondent 13).

Numerous outreach and public relations campaigns were developed to convey messages about trees and the campaign to New York City residents. NYRP worked with professional design firms to create logos, identities, and various branding approaches to the campaign. This included public service announcements, advertisements on bus shelters and subways, tree tags and signs, and giveaway shirts and buttons for volunteers. A large-scale print ad campaign from MillionTreesNYC featured images of trees and people in the city, touting trees as "Zen masters: (Trees do more than you think. They promote relaxation and fitness, enhance our emotional and mental health, and even encourage us to drive a little slower)" and "exercise partners: (Trees do more than you think. While protecting us from the sun, they encourage outdoor play and exercise—helping in our fight against obesity)." The campaign also effectively used visual imagery, such as renderings of city blocks before and after blockwide tree planting, to demonstrate the transformative aesthetic effects of trees on the streetscape. A later marketing effort stripped down the message to a much narrower focus on personal commitment, through the use of the slogan "I'm In." This marketing approach continued until the completion of the campaign; at the celebratory, final planting day, tee shirts stating, "I'm the Millionth" were distributed to attendees.

Campaign leaders and their allies described the challenges of using marketing to move from raising awareness, to changing perceptions, to cultivating behavioral changes. Outreach was successful in getting the word out and making MillionTreesNYC one of the most visible efforts of PlaNYC. Yet one interviewee cautioned against overreliance on a marketing approach as opposed to building a truly grassroots movement, saying: "[Be] careful going too far down this marketing road or putting all your eggs in that basket because then you find that . . . the *soul* of the actual movement is gone. And you're just propping it up with these great images or fun slogans or this and that" (respondent 28). The campaign was a delicate balancing act between efforts like the marketing campaign that reached millions of people citywide in a somewhat fleeting way and more in-depth programmatic engagements that met fewer people in a more lasting way, such as education and stewardship initiatives.

Finally, although not foregrounded in the public messaging about the campaign, managers brought a sense of distributional justice to the implementation of the campaign. DPR officials were cognizant of the way in which the urban forest had developed unevenly over the course of the past several decades and saw this large-scale campaign as a chance to correct this inequality, aiming to plant every available street tree planting location, beginning first with the areas that were most lacking in trees. Even when managers did not use the language of justice, they sought a defensible rationale that they could present to the public for why they planted in what places and in what sequence. Numerous times, public officials whom I interviewed cited the importance of being accountable to the public and of being fair and ethical in resource allocation decisions. Additionally, the belief that everyone deserves access to clean, green neighborhoods was one of the driving forces behind Midler's creation of NYRP. As a result, the organization was committed to greening in "high need neighborhoods" (respondent 10). Finally, this same focus on distributional equity across geographic space was also present in the most visible open space goal of PlaNYC: that all New Yorkers should be within a ten-minute walk of a park. One respondent reflected on PlaNYC's progression from goals related to attracting elites, to goals serving all residents:

> [PlaNYC] starts out with an elitist idea to a certain extent that says, "Well, our streets ought to be attractive, we want our property values to go up, and we know . . . about the property value impact of street trees." . . . But then when you look at it and . . . we started doing the math [with DPR] about where there were and were not street trees, it became clear that this was an initiative that was actually as much or more about environmental justice as it was about creating elite property values. And so once we set the goal that says, "Look, every place that it is feasible to put a sidewalk tree, we would like to put a sidewalk tree" you immediately have a policy that fills in the valleys. And, frankly, there aren't that many places on the Upper East Side that you can put more street trees, but there are lots of places in the South Bronx. And so it was one of these things that turned a transition from a hard infrastructure plan into a sustainability plan, and an elitist, global competitive story into a quality of life-for-all story. (respondent 49)

Thus, tree planting was a powerful and flexible tool that could help advance aims of the competitive city but could also address some of the prior inequities in the local environments of New York City residents.

Harnessing the Network and Engaging the Community: Advisory Committee and Public Programs

Realizing that MillionTreesNYC would be prominent in the organizational landscape of New York City's environmental groups, its leaders deliberately crafted roles for allies—and even potential competitors—to take part in the campaign. The MillionTreesNYC advisory committee institutionalized a network of advisors of approximately four hundred individual members from more than one hundred organizations. New York City's environmental sector is highly diverse and complex, with numerous public, civic, and private entities involved (Fisher et al. 2012; Connolly et al. 2013).[2] Prior efforts to convene local environmental groups in New York City, such as attempts to create a Nature Network in the early 2000s, had not met with a great degree of success. One respondent described New York City's existing "baggage of people, personalities, organizations, and events" as something that had to be reckoned with in the creation of this campaign (respondent 28). Thus, it is noteworthy that—at least for a window of time during the formation and early stages of the campaign—MillionTreesNYC became a focal point for the active engagement of numerous environmental organizations through its advisory committee. Given the momentum of PlaNYC and the enthusiasm of Bloomberg and Midler, the campaign was seen as the "train leaving the station," with numerous organizations and individuals eager to shape the effort as it unfolded.

The campaign had varied effects on the citywide civic stewardship broker organizations identified in chapter 1. Groups working on urban forestry issues generally lauded the effort, noting that MillionTreesNYC raised public visibility of their programs—so rather than competing for a limited pool of resources and public attention, the campaign helped to "raise all boats." As such, positive effects ranged from increasing budgets, to expanding the scope of the organization, to attracting additional volunteers. Many of these organizations became involved in the implementation of the campaign: by serving on the advisory committee; by partnering in the recruitment of volunteers for large-scale citywide volunteer reforestation days; or by serving as partners in the MillionTreesNYC Stewardship Corps. Other groups were less enthusiastic and offered a range of criticisms about MillionTreesNYC. Some groups had assumed that the campaign would bring new resources to their own organizations, but these did not materialize. Others focused on the overall scope and intent of the campaign—noting that planting one million more trees in New York City might be the wrong goal to pursue and that there might be other more pressing issues on the sustainability agenda. Expanding on this, others argued that the campaign was largely about public relations and was nothing more than a "window dressing" on the city. Others critiqued the implementation of the campaign, noting that the stewardship component was not fully thought through at the outset. Although the goal of the campaign was to "plant and *care for* one million trees," some civic environmental groups felt that the leadership of the initiative did not place stewardship front and center in the campaign.

The advisory committee offered a twofold benefit to the campaign of providing a means for outside experts and civic stewardship groups to contribute input to the campaign, but also heading off these sorts of critiques before they emerged. By being asked to participate, potential critics or competitors would feel invested in the campaign. Indeed, there was some concern by campaign leadership in the face of external critique in the early days of the campaign.[3] In addition, latent tension existed between other greening nonprofits and NYRP, which the advisory committee hoped to help bridge. One leader of the campaign said, "I'm not sure too many people own the initiative outside of NYRP. Because people said, 'What's in it for us? Why should we do it when we're not getting any money?'" (respondent 35). Campaign leaders acknowledged the procedural importance of the committee in light of prior critiques of PlaNYC as top-down. The committee was also to serve as a source of ideas, skills, and resources for the campaign. One respondent noted a desire not to "reinvent the wheel" with this campaign, and to build on the existing expertise and investments of dozens of groups citywide. The broad membership aimed to support the longevity of the campaign beyond the 2014 change in mayoral administration. Campaign leaders used the language of creating a "movement" around urban forestry, wherein "this campaign was larger than the Parks Department and NYRP alone" (respondent 15). Another official echoed the language of movement-building: "I think the advisory board serves several purposes . . . breadth and also longevity. So, the advisory board allows us to be able to say in a very real way that it's not just about tree planting; that we want MillionTreesNYC to be about creating a whole . . . urban forestry movement for planting and care and awareness. . . . So when the next administration comes . . . this is too popular to just say, 'Okay sorry. We're going to stop at 600,000'" (respondent 4). The list of participating organizations in the advisory committee was published on the MillionTreesNYC website and served as a sort of imprimatur for the campaign.

Although the intention of the advisory committee was to foster meaningful participation and collaboration, it did not always live up to this ideal. While some valued the opportunity to give input to the campaign and felt listened to, others argued that the input felt token or that outside expertise was disregarded, as annual meetings took on the structure of reporting out from the subcommittees as opposed to actually generating discussion or debate. Some critiques came from within MillionTreesNYC, with one leader expressing questions as to who was advising whom: "Are they advising us or are we advising them? Are we telling them what to do? Are they telling us what to do? . . . The relationship between the advisory committee and Parks and NYRP is very unclear" (respondent 11). Finally, there was a general acknowledgment that it was challenging to sustain interest and engagement in the long term, over time, among volunteer committee members whose core job was not the MillionTreesNYC campaign.

Topical subcommittees were formed in the following areas: stewardship and education (which was later subdivided into two groups); research and evaluation; public policy initiatives; marketing and public relations; and tree planting. Green jobs was added later. The structure of each subcommittee's leadership included representatives from DPR, NYRP, and an outside organization to ensure broad participation. The level of engagement of outside entities, serving in a volunteer capacity, waxed and waned over time depending on organizational circumstance and in response to the unique challenges and opportunities each subcommittee faced. One respondent acknowledged the varying levels of engagement of different committee members: "There's always going to be great lists of organizations and only five percent are really going to do the work" (respondent 28). Programs in *education, stewardship, green jobs,* and *research* helped expand the reach and scope of the campaign, with varying degrees of success.

Education

Environmental education—particularly for youth—was seen as one of the main programmatic efforts that NYRP, DPR, and other partners could leverage to enhance public understanding of urban trees and to help build a broad-based urban forestry movement. NYRP had experience with developing educational programs, and this was cited as one of the reasons that they were an apt partner for the campaign. A nonprofit employee said, "It wasn't just putting trees in the ground . . . there was much consideration given to having an educational component, marketing component, and a communication strategy around it" (respondent 35). Similarly, for DPR the campaign presented an opportunity to actively engage the public, instead of simply "planting a lot of trees" (respondent 27). DPR had several longstanding environmental education programs that it could both leverage and expand through this campaign. DPR's Urban Park Rangers had already been in existence for more than thirty years and offered school programs and mentoring opportunities for New York City youth as well as nature walks and interpretations for all ages.

Despite this potential, one respondent noted that MillionTreesNYC failed to capitalize fully on a huge opportunity to emphasize environmental education in tandem with tree planting on or near *school grounds,* particularly given the 1.1 million children in the New York City public school system—the largest in the country (NYC DOE 2012). NYRP created a new curriculum called RESPECTree, and DPR expanded its Urban Park Rangers' "The Natural Classroom" program with a new program entitled "TreesNYC: Something Big Is Taking Root." However, due to staffing, budget, and time limitations, RESPECTree was delivered only to ten schools per semester, and no large-scale citywide school planting program with integrated curriculum was developed by NYRP. As the campaign progressed, NYRP staff continued to develop additional educational curricula and programs, including "15 For Trees"—a fifteen-minute lesson plan appropriate for grades K–8—and "Make Every Day an Arbor Day"—a workbook of activities intended to be linked to a tree planting. Nonetheless, one interviewee argued that the educational efforts were subsumed by a greater emphasis on visually changing the landscape, noting: "I don't think Bette thought about the other stuff like public education and stewardship. It was about beautification" (respondent 35). Both the Bloomberg administration's emphasis on tree planting metrics and Midler's emphasis on planting-as-beautification failed to place educational programming front and center to this campaign. In contrast, stewardship programs delivered by DPR and NYRP continued to evolve and deepen over the course of the campaign.

Beyond the programs delivered by the lead partners, the education subcommittee sought to catalog all the other tree-related environmental education programs that existed in the city. Harnessing existing resources was an attempt to prevent duplication of effort. The campaign developed an online and print educator toolkit of available curricula from which educators could pick and choose. After that initial stage, the subcommittee sought to track metrics on how many students and teachers participated in various tree planting curricula and programs. This proved a challenge for a volunteer committee with no incentive or remuneration for reporting on these metrics. Finally, in later stages of the campaign, formal educational partnerships between MillionTreesNYC and large organizations—such as the Girl and Boy Scouts and the "Green Points Challenge" with Recyclebank—continued to emerge.

Stewardship

The challenges of tree maintenance and the opportunities for public stewardship drove the development of the stewardship subcommittee, partnerships with other service and volunteer organizations, and eventually the creation of the formal programs StewCorps and TreeLC. Much as the education committee did in the early days of the campaign, the stewardship committee consulted experts in stewardship, via participatory fora, about best practices and guidance for the campaign. The need to formalize collaboration on stewardship in an ongoing way was quickly identified—particularly because so much of the success of the campaign as a planting effort depended upon the ongoing survival and thriving of the expanded urban forest. Initially, this took the form of partnering with outside groups such as NY Cares (a nonprofit) and NYC Service (a division of the mayor's office) to recruit volunteers and organize large-scale planting days. These large-scale planting days in natural areas continued to be organized by DPR staff throughout the life of the campaign. Each spring and fall, thousands of volunteers planted even larger numbers of trees in a few coordinated hours of planting at sites across the city.

With funding support from the Mayor's Fund, StewCorps was developed to help grow engagement in street tree stewardship. It started in the summer of 2009 as a partnership between MillionTreesNYC, DPR's GreenThumb community gardening division, the nonprofit urban forestry group Trees New York, the major botanical gardens in the city—New York Botanical Garden (NYBG) in the Bronx, Brooklyn Botanic Garden (BBG), and Queens Botanical Garden—and DPR's Staten Island Greenbelt. Each of these primary StewCorps partners agreed to deliver a set number of tree care and stewardship workshops and activities in exchange for a grant from MillionTreesNYC in the $10,000–25,000/year range, depending on the number of people served. Although both NYRP and DPR had experience in cultivating stewardship, this program acknowledged the long track record of these other prominent greening groups in this area. This resource sharing showed a level of collaborative governance with outside entities beyond that which occurred on the education subcommittee.

To foster innovation, the StewCorps partners were given latitude to customize their training and programming around trees as best suited their organization and the audience they served. BBG's "Greenest Block in Brooklyn" competition was used as a way of engaging residents' productive sense of competitiveness. BBG argued for a terminology switch from "tree pit" to "tree bed" as a rhetorical device for getting people to think about the area surrounding the tree as a planting area that deserved attention and care. This language of the "tree bed" and many of the gardening tips for planting in those sites were taken up and incorporated into MillionTreesNYC stewardship trainings. Trees New York delivered its Citizen Pruner curriculum that it had developed and honed since the 1970s, expanding its reach and scope to serve more people and groups with MillionTreesNYC funding. GreenThumb developed interactive games about trees in order to engage community gardeners in fun and informative ways.[4]

Stewardship programming continued to evolve and expand throughout the later stages of the campaign, with street tree stewardship organized via the

TreeLC program of MillionTreesNYC and natural areas stewardship supported via the Natural Areas Volunteers of DPR. TreeLC consisted of delivering public workshops, training "care captains" via a train-the-trainer model, offering stewardship tips via an online TreeLC manual, and offering competitive mini-grants to community-based stewardship groups. In 2012, TreeLC delivered 172 workshops to 3,405 attendees (Campbell and Monaco 2014). The train-the-trainer and mini-grant model was designed to expand the reach of the program beyond what MillionTreesNYC and DPR staff could do, acknowledging the advanced level of stewardship of some community-based groups and the importance of grassroots-led, bottom-up organizing. The Natural Areas Volunteers encouraged residents to get involved with natural areas not only as volunteer planters but also as ongoing stewards of these sites. This was an important shift, as previously the public was able to interact with forests only during one-time planting days; new opportunities were created for longer-term and sustained engagement with these sites. These volunteers are required to attend training sessions, are supervised by DPR staff, and can be further trained and certified as "NAVigators" with the ability to adopt a location if they commit to twelve hours of work. TreeLC and the Natural Area Volunteers programs reflect the maturation of stewardship programming over the course of the campaign—with phases of information gathering, experimentation with different networked relationships, and consolidation of programming under DPR leadership.[5]

Green Jobs

Launched in winter 2008, the MillionTreesNYC Training Program (MTTP) reflected the zeitgeist of the recession years and of the Obama administration, with its focus on training, green jobs, and employment. The program offered paid job training in urban environmental restoration work for low-income, eighteen- to twenty-four-year-old adults who were previously disconnected from the workforce (Maddox et al. 2010). This effort received support from the NYC Center for Economic Opportunity—a mayoral antipoverty initiative—and from the Altman Foundation, the Arthur Ross Foundation, the Dodge Foundation, and Bloomberg and Rockefeller funding via the Mayor's Fund (MillionTreesNYC 2009).

Over five seasons of operation, the program graduated 104 trainees from a seven-month job-training program in which they were paid while learning the skills of horticulture, arboriculture, and ecological restoration (MillionTreesNYC 2009). In addition, trainees received environmental education and certification from affiliate organizations like NYBG and BBG, as well as basic job skills like commercial driver certification, and "soft skills" in time management, office behavior, and résumé writing. Trainees were also provided with mentors and a case manager (Maddox et al. 2010). In 2009, graduates of the first MTTP class were assisted with placement in one-year follow-up jobs with local agencies and nonprofits, using $2 million in funding from the U.S. Forest Service to cover those placements. A study of MTTP graduates found that, despite challenges common to this population, "Green job training and employment present real opportunities for intellectual stimulation and an increased sense of accomplishment, due in part to the uniqueness of environmental work. Individuals reported positive environmental attitudes and behaviors as a result of green jobs training and employment" (Falxa-Raymond et al. 2013).

Despite the demonstrated benefits of the program and managers' support for the idea of green jobs, several interviewees offered concerns about MTTP. One questioned the placement of MTTP graduates in the workforce over other qualified candidates, citing that they still were not prepared for full-time jobs alongside professionals. Others were concerned with the high costs of the program and the challenges in keeping it financially sustainable during the economic downturn without ongoing federal or foundation support. Once the Altman and Ross funding was exhausted and not renewed, in the fall of 2011, the program had to downscale the number of trainees that it could serve, and the training track that was administered by NYRP was cut, so that only DPR was administering the program via its Green Apple Corps. The job placement program lasted only as long as the USDA funding was available. While the campaign tried to stretch into this new terrain, it was not sustained alongside the core functions of tree planting and maintenance. Green jobs were only a priority program for a narrow window of time following the global financial crisis.

Research

The level of engagement of the research subcommittee was the highest and most consistently sustained of all the subcommittees. This group was unique in that it had the institutional support of the New York City Urban Field Station, a joint effort of DPR and the U.S. Forest Service, focused on cultivating collaborative research networks on urban social-ecological systems. Moreover, numerous academic researchers at New York University, the New School, Columbia University, Cornell University, and others engaged with MillionTreesNYC as a research topic. Respondents noted the productive self-interest of researchers interested in accessing data and field sites, conducting interviews with managers and stewards, and publishing articles about MillionTreesNYC. Academics felt that the transformative change to New York City's landscape, ecology, and environmental governance brought about by the campaign merited study.

The research group served as a gathering space and forum for interested researchers and had a core aim of strengthening the research-practice interface (Campbell and Monaco 2014). It held a preliminary brainstorming meeting of approximately fifty people in October 2008, discussing the role of research in the MillionTreesNYC campaign. Following that, subcommittee members, with guidance from the U.S. Forest Service, created an urban forestry bibliography free for download, in an attempt at synthesizing some of the existing research in the field. This was a similar step to the toolkit created by the education subcommittee—assessing and synthesizing the existing knowledge and work in the field. Next, the research group organized a two-day workshop of meetings and field tours of planting and stewardship sites in the spring of 2009 that convened a larger set of researchers and led to a report about the effort to integrate research and management practices, including identifying key knowledge gaps and research questions (MillionTreesNYC Advisory Committee 2009). Then the group curated a two-day research symposium of invited speakers, submitted talks, and poster sessions at the New School in the spring of 2010 that was attended by more than two hundred people. It led to a special issue of the online journal *Cities and the Environment,* which included twelve research articles and twelve posters from the conference—including publications with both researchers and park managers as coauthors (Svendsen and Lu 2010). The symposium was a significant milestone in helping to create collaborative networks of people engaged in MillionTreesNYC and to advance the co-production of knowledge (Campbell et al. 2016).

Following these two major gatherings, the subcommittee shifted away from its role as a convener. Instead, individual research and evaluation projects proceeded independently, albeit sometimes with interconnections that were facilitated by the network of colleagues fostered through the subcommittee. For example, a subset of committee members and outside partners began National Science Foundation–funded research on changes in stewardship in New York City over twenty-five years (Connolly et al. 2013; Locke et al. 2014). Others conducted field ecology research or "designed experiments" on forest restoration initiatives (McPhearson et al. 2010; Simmons 2010; Felson et al. 2013; Falxa-Raymond et al. 2014). Still others examined community engagement, stewardship, volunteerism, and green jobs (Moskell et al. 2010; Fisher et al. 2010, 2011, 2015; Falxa-Raymond et al. 2013). Members of the subcommittee reconvened for a final effort at synthesizing research and practice in the MillionTreesNYC campaign in the form of a print and online report that was distributed nationwide via urban forestry networks in order to help inform other campaigns emerging in cities across the globe (Campbell and Monaco 2014).

Change over Time: Challenges, Opportunities, and Maturation

A multiyear urban forestry campaign is not a static thing. It matures, learns, and experiences setbacks and windfalls. Given this, what were some of the key turning points experienced in the campaign? External forces, particularly the 2008 global financial crisis, as well as internal forces, such as leadership changes, played a role in those shifts. Despite the potential to be a salient window for policy change, the update to PlaNYC in 2011 had little impact on the implementation of an already-existing mayoral campaign. Finally, what happened to this Bloomberg-led initiative under his mayoral successor, Bill de Blasio?

Financial Crisis, Economic Downturn, and Budget Cuts

Although the campaign constantly evolved throughout its implementation, perhaps the single greatest turning point it experienced was the financial crisis of 2008. All city agencies, including DPR, sustained 30 percent budget cuts in fiscal year 2009. This municipal belt-tightening and emphasis on fiscal discipline continued the sorts of neoliberal measures that have been used in New York City since the 1970s crisis (Shefter 1985; Brecher et al. 1993; Brash 2011). It is not uncommon for the DPR budget to be one of the first cut in difficult economic times, as it is seen, as one leader said, as "soft stuff"—more politically malleable than the school, police, or fire budgets (respondent 27). Another decision-maker corroborated:

> Nobody hates parks. Most people would say that they are important, a quality-of-life amenity. People love them. They have exercise, cultural events, education. All these wonderful things happen there. But it never translates into budget support. I think both the public and the elected officials sort of take it for granted that if the funding is reduced, nobody's going to die as a result. Either the grass won't get cut or the garbage won't get picked up as frequently, but it's not going to hurt the city immeasurably or permanently. (respondent 47)

These cuts led to an agency-wide hiring freeze at DPR in 2009, which meant that even when existing employees left for other jobs or new locations, there was difficulty in filling their positions in order to fully staff divisions. In particular, this had implications for CFH and NRG as they faced leadership changes and subsequent reorganizations in response to those changes.

Although cuts in various program areas occurred, the capital dollars allocated via PlaNYC for tree planting were *not* lost in these 2008–2009 budget cuts. The agency made a priority of reaching tree planting targets and instead made cuts to other programs, such as the trees and sidewalks repair program, the pruning budget, and the Greenstreets program.[6] Indeed, the bureaucrats within DPR made cuts to other signature PlaNYC projects—like the regional parks initiative that targeted a few major parks citywide for renovation and construction—before affecting tree planting budgets. After the initial round of 30 percent cuts, budget cuts continued and deepened with another 8 percent cut, during which losses to the MillionTreesNYC program were sustained.

These cuts to the maintenance, pruning, and stump removal budgets occurred quietly and did *not* trigger an effective public outcry. Neither residents, nor elected officials, nor members of the advisory committee pressured the city council or the mayor's office to resist the across-the-board cuts to the maintenance budget, despite the prior calls for attention to maintenance over the course of the campaign. This reveals that while the advisory committee may have been stakeholders in the campaign, they did not have the clout to shape the higher echelons of decision-making at City Hall, OMB, or the city council. It also reflects a subtle tension that may be produced by the practices of a high profile, public-private partnership. While the municipality exercised fiscal discipline and made deep cuts to city agency operations, the visible and glossy public relations campaigns crafted by NYRP and its design consultants for MillionTreesNYC continued. According to the messages received by the public, the campaign was humming along. Yet large changes were afoot. These sorts of decisions made by bureaucrats are not always fully visible, but they are crucial to how the campaign was implemented.

At NRG reforestation sites, prior to 2008, planting was handled by a combination of in-house crews and contractors, but after the global financial collapse and subsequent municipal cuts, DPR began to depend more upon volunteers in large-scale tree plantings. Each fall and spring, several sites across the five boroughs and twenty thousand trees were planted in single-day volunteer planting events (see figure 4.1). Volunteers came from workplaces (including a large contingent of corporate employees), schools, clubs, and other civic groups (Fisher et al. 2011, 2015). Despite the significant preparatory work that was required—recruiting and managing volunteers, predrilling holes for planting, allocating trees in the correct locations—using volunteers for their physical labor still served as a cost-cutting measure compared to using paid contractors. Although cost-savings was one driving rationale, DPR also hoped that these volunteers would come to feel more invested in park sites that are often overlooked or less visible than traditional recreational sites. From the perspective of the leaders developing MillionTreesNYC as a volunteer program, active engagement with tree planting events was one of the key points of contact between the public and the campaign and a cornerstone of DPR's stewardship programs.[7]

FIGURE 4.1 Volunteers at NRG reforestation planting event at Alley Pond Park in Queens. Photo by author.

Starting in 2008, fewer street trees were planted per season due to budget cuts. While DPR's commitment to fully stock streets citywide remained, the calculation of the number of trees that could be planted on each street length was reduced—meaning trees would be planted in lower densities on streets. The original PlaNYC goal was to plant 220,000 street trees citywide; in the end, approximately 164,000 street trees were planted during the campaign. Thus, even the firmest of quantitative goals that was explicitly named in the plan was subject to market vicissitudes as well as to the very real and on-the-ground challenges of infrastructure conflicts and space limitations that exist when one moves from a planting goal to implementation. These adjustments were made within the operation of DPR, and the campaign rolled along, publicly achieving its goals. In addition, as previously mentioned, the MTTP green jobs program ended after its federal and private sector funding sources were exhausted; the city could not afford to sustain this high-cost program without these partners.

For NYRP, the financial downturn also coincided with the aging of the campaign—leading to fundraising challenges. Funders became increasingly reluctant to give due to the economy and to fatigue with the multiyear campaign. At the outset of the campaign, and building off the momentum provided by the Bloomberg and Rockefeller philanthropic donations, NYRP was able to secure corporate sponsorships. But over time, corporate funding priorities began to shift (e.g., in the case of Home Depot's giving, a change in focus from urban forestry to housing development occurred). Fundraisers struggled to position the ten-year MillionTreesNYC campaign as novel and fresh, as corporate and philanthropic donors' eyes began to wander in search of the next, newer effort. This raises the question that if trees are considered part of a city's core infrastructure, should they be funded via private dollars, corporate donations, and philanthropy?

NYRP had to continually reframe the campaign around different narratives, as well as change the emphasis in the implementation, in order to stay ahead of the challenging funding environment. The main shift in NYRP's implementation involved a decrease in the number of trees that they would plant directly. As their responsibility for the number of trees planted decreased from four hundred to three hundred thousand, they simultaneously shifted their emphasis away from "direct planting" efforts and toward tree giveaways and "influence plantings," at significantly lower cost to the organization. Tree giveaways continued as the main programmatic activity of NYRP until the end of the campaign.

Updating PlaNYC

The 2011 update to PlaNYC occurred in a very different economic and political context from that of the original plan. Because of the fiscal reality, PlaNYC was no longer powered by large infusions of capital. City Hall emphasized repeatedly that PlaNYC 2.0 (as the 2011 update was informally called) would require the city to "do more with less" (respondent 15). A municipal official described this shift from 2007 to 2011: "The 2011 update, we were under orders: no new capital commitments. No new expenditure commitments were allowed in the plan. So it's a very different fiscal circumstance for the update compared to the original, in terms of what the administration was able to commit financially. We had to be more modest in our milestones as they pertained to expenditures" (respondent 46). At the same time—and perhaps out of financial and political necessity—the city began to open the planning process to new voices. While the initial document was tightly controlled and developed in a top-down fashion, the PlaNYC 2.0 process was more inclusive. The Office of Long-Term Planning and Sustainability (OLTPS) hired coordinators whose primary job was to liaise with community residents and organizations in order to gather their ideas, engage them in dialogue, and help develop a broader network of supporters for PlaNYC. Borough-based public hearings were held, as well as topically focused meetings at City Hall and other public venues.

Despite its broader inclusiveness, PlaNYC 2.0 was seen by many agency officials as an exercise with which they had to comply, rather than a strategic process empowered by strong top-down support from City Hall. PlaNYC 2.0 had little effect on urban forestry. Almost all interviewees reported a lack of engagement in the process and almost no impact of the plan's issuance on the practices of the campaign. One interviewee called it "just a legal requirement" (respondent 47); others had not even read or circulated the plan among their staff. Indeed, without the infusion of funding, it was noted that the mayor's office and OLTPS had few tools for reforming or changing agency practices outside of budgetary allocations. PlaNYC 2.0 was seen primarily as an administrative "box to check," as opposed to a political opportunity. Thus, there was a disconnect between a more broadly inclusive structure for formulating goals and a lessened ability for goals to be implemented.

In contrast, some of the partners in the MillionTreesNYC campaign felt that PlaNYC 2.0 went in a positive direction by drawing public attention to MillionTreesNYC as a successful initiative. Despite the lack of new funding commitments, many respondents appreciated the broader inclusiveness of its process, the references to stewardship and maintenance of existing assets (parks and trees), and the recognition of the role of civic groups and volunteers. In fact, one interviewee argued that the way in which MillionTreesNYC was implemented, with heavy volunteer involvement, its StewCorps partners, and a broad-based advisory committee, may have informed PlaNYC 2.0: "It was sort of a revalidation. PlaNYC 2.0 . . . concentrated on the citizen involvement in stewardship aspects of what we're doing, which is great . . . and community involvement has become such a salient part of many different PlaNYC initiatives. I think perhaps Million Trees has informed PlaNYC 2.0 more than vice-versa . . ." (respondent 1). Others said that OLTPS did a better job of outreach to NYCHA, including that agency as a part of PlaNYC 2.0. Finally, one interviewee felt that PlaNYC 2.0 better acknowledged the role of NRG and the importance of "natural areas" to the tree planting goal, highlighting this more than the heavy emphasis on street trees in the first edition of the plan. Still, this attention in the document did not amount to any additional political or financial support.

Leadership Transitions and the End of the Campaign

Top-down leadership played a crucial role in the conception and implementation of the campaign. When the initial leaders left or changed positions, ripple effects were felt throughout the network. First and foremost, this campaign was a mayoral priority, which drove the substantial budgetary commitment, aggressive timeline, obsessive reporting and tracking, and concerns over what would happen in future mayoral administrations. This campaign was not simply urban forestry business as usual. Although the campaign was initially created to last through 2017, its leaders made a concerted effort to finish as much planting as possible before the 2014 change in mayoral administration and to plant the full million trees by 2015. Despite being a long-term sustainability plan with time horizons through 2030, PlaNYC was entirely an executive initiative, wholly identified with the Bloomberg administration. Midway through the campaign, one city bureaucrat said, "One of the things that you quickly learn in politics is that the succeeding mayor has no stake whatsoever in burnishing the legacy of his predecessors. . . . I would certainly hope that everything gets done before 2013, because I suspect that what doesn't almost certainly ain't going to happen" (respondent 41). Over the course of Bloomberg's final term, MillionTreesNYC leaders worked to build a sense of inevitability such that the future mayor would have been ill-advised to quash the effort. In addition, NYRP, which does not face any city administration time limits, provided some potential continuity for the campaign. However, the nonprofit underwent its own leadership changes at all levels from executive director, to MillionTreesNYC director, to numerous changes in field staff.[8]

MillionTreesNYC continued on its own momentum in 2014 and 2015, the first two years of de Blasio's term, relatively outside of the mayor's core areas of focus. During the administration change, a number of key leaders of MillionTreesNYC at DPR left to seek other positions, advisory committee meetings ceased, and a sense that the winds had shifted permeated the campaign. The de Blasio administration did not focus as overtly on environmental issues and instead placed social justice front and center. His signature initiative of his first year was the rollout of universal, public prekindergarten. His policies with DPR focused primarily on an equity plan to redistribute capital funds to under-resourced neighborhood parks, called the Community Parks Initiative, giving much less attention to forestry, tree planting, and natural areas. Despite this internal reorganization at City Hall, large-scale volunteer planting events continued to be held, and stewardship workshops and mini-grants were delivered, all of which continued apace as part of DPR's forestry operation. OLTPS put out the required PlaNYC 2014 progress report, with an accounting on the existing programs, including MillionTreesNYC. The report stated that the campaign was "27% ahead of schedule, having planted 830,000 trees and held numerous events to teach New Yorkers how to care for them" (City of New York 2014a, 9). Reflecting the maturation of the campaign, many of the metrics reported in the tables emphasized the degree of public engagement and volunteerism associated with the campaign, as well as the ongoing commitment to improving tree survival.

Midway through the campaign (around 2011), my interviewees speculated about its future. When asked what NYRP would do with MillionTreesNYC going forward, a respondent said:

> I think we'll keep doing tree giveaways. . . . I think that we'll continue to plant as much as possible. . . . I don't think it matters [if we don't reach a million trees]. I think that people appreciate honesty. . . . I think you can say to people, "This is a very ambitious project. We've worked tirelessly. These are all of our accomplishments over the last few years. . . . However, given the financial challenges . . . we're not . . . going to complete the project in the targeted timeframe. However, Million Trees is not a ten year initiative. It's a concept. It's an idea. And so, please support us and really work with us to fulfill this project. Plant trees in your home, outside, in your yard. Work with your schools if there's lawn space on the schools. Find places." And, maybe if we created some best practice guides that were like, "This is what we learned from the initiative. This is what we had done well. We can support you with these resources, non-financial resources to help plant more trees in your community" . . . I think people would respond to that. (respondent 55)

Similarly, another respondent offered that the campaign could have shifted the focus away from reaching the one million tree mark: "I mean, at the end of the day everybody was excited about the initiative and saw the potential. . . . But at the end of the day—if we only at the end of the initiative get 800,000 trees in the ground, it's still a success. Like, no one has failed because we got people excited about tree planting. More people are aware of it. And we got 800,000 more trees in the ground . . . it's a no lose proposition" (respondent 35).

However, the identity of the campaign was so tightly connected to the MillionTreesNYC brand that this shift did *not* occur. Instead, DPR pivoted the emphasis of the initiative from aggressive, municipally led tree planting to a broader movement around civic stewardship of trees, and NYRP scaled back direct planting to focus on tree giveaways. In the later stages of the campaign, MillionTreesNYC and DPR staff and their partner organizations worked to build a corps of committed volunteers and engaged residents who would help carry forth the initiative via the TreeLC program and the Natural Areas Volunteers. DPR continued to embed stewardship more fully into its operations, and in late 2014 they created the new position of director of stewardship. The director would work across site types (street trees, "natural area" forests, and eventually wetlands) to engage the public in caring for nature in the city. They finely honed their volunteer planting experience, created mass one-day planting events each fall and spring, and continued to develop a staff focused entirely on stewardship working under that director. As of July 22, 2013, the www.milliontreesnyc.org website counter showed 757,386 trees planted; on October 28, 2014, it read 912,524; and on November 20, 2015, the successful completion of the campaign was celebrated in a public tree planting featuring Bloomberg, de Blasio, other elected officials, Midler, and campaign supporters and volunteers.

5

GROWING IN THE CITY

From Community Gardening to Urban Agriculture

There is a complex network of political actors, narrative storylines, and material practices involved in creating, advocating for (or resisting), and maintaining urban agriculture in New York City. These networks extend laterally to other sectors, geographically outward in space, and historically through time. Understanding contemporary networks requires examining a brief history of community gardening as a social movement in New York City since the 1970s, including the 1990s garden crisis. Since those turning points, there has been a recent renewed wave of interest in urban agriculture and local food systems dating to the mid-2000s. This more recent trend can be parsed into its various threads, which range from a focus on local food production, to commitments to food justice, to an interest in neighborhood stabilization and youth empowerment. Taking into account numerous actors and their relationships, as well as the multiple, sometimes overlapping, and sometimes competing storylines that they deploy, helps us to understand urban agriculture as a political and discursive process—in addition to being a fundamentally material practice of growing crops and managing land in the city. In particular, new constituencies focused on transforming local and regional food systems have emerged, thereby influencing the way in which urban agricultural production at gardens and farms is framed.

Community gardening and urban agriculture are not synonymous. Although community gardens can be important agricultural sites (Farming Concrete 2011; Gittleman et al. 2012), certainly not all gardens focus on food production

(Ackerman 2011; Cohen et al. 2012). Instead, community gardens are *community-managed open space*. Thus, they can be space for recreation, performance, food production, gathering, cultural practices, or many other functions (Schmelzkopf 1995; Kingsley et al. 2009; Ohmer et al. 2009; Cohen et al. 2012; Mees and Stone 2012). Since the fiscal crisis of the 1970s, New York City has had one of the largest and most robust community gardening programs in the world, with a broad base of resident engagement (Von Hassel 2002; Lawson 2005; Stone 2009; GreenThumb 2010). That base was mobilized with particular urgency in the mid-1990s when then-mayor Rudolph Giuliani threatened to auction hundreds of garden sites for housing development. Legacies of the garden preservation struggle reverberate to the present day. The complexity of land jurisdictions and institutions serving gardeners continues to shape how gardens in New York City function.

Across almost all of my interviews and numerous media accounts, there was a sense that *something* is different about the current efforts in urban agriculture. While impossible to pinpoint the exact start, there is a new, growing wave of engagement in urban agriculture—and a broader interest in localizing food systems, coalescing in the 2000s. New York City has wide-ranging innovations in agricultural practices, including new rooftop farms, school gardens/greenhouses, backyard chickens, and beekeeping—all of which are rendered more visible through greater media attention surrounding urban farming and local food production. One notable difference from prior urban agricultural production is that some of these new models (both for-profit and nonprofit) are entrepreneurial, selling their produce rather than dividing it among members or donating it. Participants perceive themselves as part of a local food movement that is increasingly recognized. Media attention, high-level support and endorsement, and the economic downturn have all added fuel to a vibrant movement that grows daily. Yet this movement is far from uniform. Activists variously deploy and wrestle with critiques such as "This is not Detroit" and "You can't feed New York City through urban agriculture." From the common trope, to the underlying tensions, to the abiding themes—how is urban agriculture advancing and being contested in New York City, and what is the role of community-managed spaces within that movement?

Community Gardening in NYC from the 1970s to the Present

There have been numerous cycles of interest in farming and gardening in the city, as Laura Lawson clearly argues in her book *City Bountiful* (2005), wherein she traces this legacy to the progressive era of the 1890s in America. Often these upswings in urban agriculture correspond to crises or downswings in the economy. A long-time urban agriculturalist I interviewed identified the Depression, World War II, and the fiscal crisis of the 1970s as the most notable periods of gardening in New York City. Recent policy documents hearken back to the history of Kings (Brooklyn) and Queens Counties as agricultural production sites in the 1800s and early 1900s, lay claim to the history of Victory Gardens and school gardens, and celebrate the tradition of neighborhood stabilization and self-help through community gardening that began in response to 1970s disinvestment (NYC Council 2010; Ackerman 2011). Current activists in urban agriculture note that many of the models being developed today were experimented with in the 1970s.

The 1970s was a pivotal era not only for individual and grassroots engagement in community gardening, but also for the development of civic organizations to help support that work. These organizations remain active and relevant in current environmental networks, with many of these groups serving as the civic broker organizations identified in chapter 1. For example, the nonprofit Green Guerillas currently offers information and community organizing to gardeners citywide. Liz Christy founded the organization in 1973 (incorporated as a nonprofit in 1976) through her gardening and advocacy work in the Lower East Side of Manhattan. In this same era, the Council on the Environment for New York City (now GrowNYC) formed to address the inability of local government to maintain the city's greenspaces. Later, in the 1980s, the major botanical gardens of New York City also created programs to support community gardening and neighborhood beautification, including the NYBG's Bronx GreenUp (founded in 1988) and the BBG's Making Brooklyn Bloom event (started in 1982) and later the Brooklyn Greenbridge community outreach program.

Municipal support for gardening began in 1978 with the formation of GreenThumb, which is now a program of DPR. Originally housed under the Department of General Services (now called the Department of City Administrative Services, or DCAS), GreenThumb was a response to the city's fiscal crisis and was described by one interviewee as "an accident of history" (respondent 6). The financial collapse led to property abandonment and arson by private owners and left the city struggling to manage vacant land. The program was developed, in effect, to enable people to help themselves by offering temporary, low-cost leases to residents engaging in gardening on vacant lots. It receives federal funding through a Housing and Urban Development (HUD) Community Development Block Grant (CDBG), which allocates support to community development programs in low-income areas (Lawson 2005; Mees and Stone 2012).

Garden Crisis of the 1990s and Its Legacy

The community gardening movement in New York City was also fundamentally shaped by its shared experience of the late-1990s garden crisis. GreenThumb's long-term lease program ended in 1995 and was replaced with a system of license agreements (Lawson 2005, 260). Starting in 1998, Giuliani targeted community gardens as potential sites for housing development. He transferred a number of sites from DPR's jurisdiction to Housing Preservation and Development (HPD), the city's affordable housing development agency.[1] In May 1999, the city placed 113 gardens on unrestricted auction, to go to the highest bidder (Lawson 2005, 261). The bulldozing and imminent auction of these gardens led to large-scale protests by gardeners and their allies, who rallied at City Hall and built encampments in threatened garden sites to attract attention from the media (see figure 5.1) (Fox et al. 1985; Schmelzkopf 2002; Von Hassel 2002; Lawson 2005; Stone 2009; NYCCGC 2010; Mees and Stone 2012). Older organizations like Green Guerillas, as well as newer organizations formed specifically in response to this threat, such as More Gardens!, helped organize and lead these protests. Activists from the Brooklyn Alliance of Neighborhood Gardens sought to counter the narrative that housing and gardens existed in a one-to-one tradeoff, issuing a postcard stating, "11,000 vacant lots in the city, 500 community gardens" (NOSC 2000). Overall, the community garden crisis—as well as the social movement that mobilized in response to the threat—became a

FIGURE 5.1 Community gardeners and allies on the steps of City Hall protesting the bulldozing and auction of community gardens in 1999. Photo courtesy of Edie Stone.

local, national, and even international news story that has also been analyzed and discussed for decades by activists, other practitioners, and scholars.

In the final hours, the New York State Supreme Court stopped the unrestricted auction, required a review of the potential environmental impact of the sale, and issued a "cease and desist" order to stop the imminent development of garden sites (NYCCGC 2010). This provided a window in which two nonprofits negotiated a purchase of numerous threatened garden sites. The Trust for Public Land (TPL) bought sixty-two gardens, and NYRP bought fifty-two gardens, for a total price of $4.2 million (Lawson 2005, 262).[2] A memorandum of agreement was created by then–attorney general Elliot Spitzer and the Corporation Counsel of New York City that created specific lists of garden parcels in different categories of protection, management, and use.[3] The terms of this agreement were carefully crafted by city bureaucrats from DPR and HPD, under constant pressure and scrutiny from garden and housing advocates.

At present, community gardens in New York City are located on many different land jurisdictions, including public land managed by DPR, HPD, Department of Transportation (DOT); public housing grounds of the New York City Housing Authority (NYCHA); and land trusts owned by NYRP and TPL. Because of this diversity, gardens are subject to differing institutional structures and rules and have access to different resources. Community gardens registered with GreenThumb (which can include gardens on non-DPR land) have the ability to set their own bylaws and governance structures, while adhering to certain standards of safety and public access. As of 2011, GreenThumb gardens must have a group of at least ten active members, must have a process by which individuals can apply to be members, and must have the garden gate open a minimum of twenty hours/week (ten hours/week of which must be posted publicly) from April through October (GreenThumb 2011, 9).

Because of its history and its unique funding source, GreenThumb has operated somewhat outside of more traditional DPR operations. GreenThumb's budget is subject to the vicissitudes of the CDBG funding levels, which can rise and fall with economic shifts and political will. As a result, the program has struggled at times to have sufficient outreach staff to serve the needs of all gardeners citywide. While in the past it had one outreach coordinator per borough, the organization had just two outreach coordinators serving the whole city as of 2011.[4] GreenThumb is not included in the Mayor's Management Report—one of the key tracking and accountability measures used to examine the performance of city agencies (City of New York 2011b). Moreover, gardens are not included in the parks rating system, which is used internally by DPR to track the maintenance and quality of public parks. While community-managed gardens would necessarily require an entirely different rating system than traditional recreational parks maintained by DPR staff, their exclusion from the ratings means that they do not receive the resources and support from DPR that other small parks receive. Instead, they are largely the product of community residents' labor and time, with a bit of baseline material support available from GreenThumb. GreenThumb not only serves gardens that are on DPR sites but also provides material resources, information, and technical assistance to gardens on all land jurisdictions—including TPL and NYRP gardens.[5] GreenThumb serves an organizing role through hosting large-scale events like the fall Harvest Fair and the spring conference, Grow Together, which has been running annually for more than thirty years. The 2012 Grow Together engaged more than six hundred gardeners in plenary talks, dozens of workshops and panels, activities, and a resource fair, serving the multiple roles of sharing knowledge, creating community, and providing technical assistance.

The fact that GreenThumb is housed within a municipal agency is something of a unique institutional structure in the community gardening sector in the United States, where many umbrella organizations that serve gardens are independent nonprofits. As part of a city agency, GreenThumb has the ability to more easily negotiate agreements with other agencies in order to provide crucial baseline services and resources to gardeners. The Department of Sanitation provided community gardens with free compost until the citywide fall leaf collection and spring compost pickup were suspended due to budget cuts in 2008 (Sustainable Flatbush 2011; DSNY 2012). The DEP has a memorandum of understanding with DPR to provide water via fire hydrant access for registered gardens. DOT can fast-track the repair of broken sidewalks around gardens. And after the immediate garden crisis was averted, DPR and HPD developed a process for transferring sites between each other in order to address the delicate balancing of housing and gardens. Staff at the two agencies can swap developable sites for use as housing with sites for use as gardens, which involves assessing both the potential for development based on size, location, and zoning as well as assessing the viability and strength of garden groups. This relationship is based on the trust, shared history, and common understanding between agency staffs—though it remains a process that is scrutinized by garden activists.

The settlement developed by the attorney general expired in September 2010, which triggered a revision of the rules governing GreenThumb and community gardens. These rules pertain to how the city deals with the development of gardens and the process of garden group license renewal, as well as what happens if a garden group becomes defunct or dysfunctional. While public officials felt that the revised rules offered sufficient protection to gardens, many activists were concerned about the long-term security of garden sites given the history of the Giuliani era (Benepe 2010). A number of organizing efforts were led by the

New York City Community Gardening Coalition (NYCCGC), Green Guerillas, and other allied groups, to ensure that gardeners knew the text of the proposed garden rules and to advocate for changes to the rules. In particular, NYCCGC organized more than three hundred gardeners to attend the August 10, 2010, public hearing on the proposed rule changes, to stage protests, and to attract media attention.[6]

Beyond trying to change the rules, some advocates suggested that gardens on city-owned land were not sufficiently protected, even when sites were under the DPR's jurisdiction. Instead, they were interested in exploring the idea of transferring gardens citywide into a conservancy or land trust, using restrictive covenants, long-term leases, or mapping as parkland. City officials countered that gardens have never been better protected through park rules, the city administrative code, and the Uniform Land Use Review Procedure (ULURP)—in part because of the institutional inertia required to change policy and procedures, stating, "The land is safer under a GreenThumb license . . . because to change out any of that under the new rules, it has to go through full ULURP. And as much as the gardeners don't believe that's a protection, they have no idea how much nobody ever wants to put anything through ULURP if they don't have to, because it takes forever" (respondent 6). Another public official corroborated the belief that the revised garden rules post-settlement offer sufficient protection such that another crisis will not occur: "My . . . sense of [the garden rules] is that I think they are way more protective than the settlement agreement was. And I think it's very unlikely that there is going to be any kind of threat to gardens like there was under Giuliani. . . . Now everybody agrees that these are valuable and that's why they all got transferred to Parks" (respondent 50). Advocates, particularly at NYCCGC, continue to push city officials for enhanced protection and permanence of garden sites citywide, while at the same time celebrating the uniquely community-based and democratic way in which individual gardens are managed.

NYCHA Gardens

Finally, gardens on NYCHA public housing grounds have an origin distinct from the DPR, HPD, NYRP, and TPL sites. Consisting of 334 developments, incorporating 2,600 acres of open space, and housing more than half a million people citywide, NYCHA is the largest public housing authority in the country (NYCHA 2012). Receiving federal funding from HUD and organized as a New York State public benefit corporation, it operates with a degree of autonomy from executive agencies. The resident gardening program began in NYCHA in the 1960s and is among the oldest community gardening programs in the country

FIGURE 5.2 Rodriguez and McKay's flower garden at NYCHA Patterson Houses in the Bronx. Photo courtesy of Lloyd Carter, NYCHA.

(Bennaton 2009). Since its origins, the program has been structured primarily around a resident garden competition and had grown to more than six hundred gardens citywide in 2011 (see figure 5.2 for a photo of a NYCHA garden). The contest was created with the idea of spurring "healthy competition" between neighbors, buildings, and developments in order to promote beautification.

NYCHA's core mission is to provide affordable housing to low- and moderate-income New Yorkers. Thus, the Gardening and Greening program (later reorganized as part of the Gardening and Sustainability unit) has often remained underresourced and out of the spotlight, with just one or two outreach coordinators, one office manager, and a few part-time, seasonal consultants serving all six hundred NYCHA gardens citywide. Moreover, there are sometimes tensions between the gardening program's aims and those of the maintenance/grounds staff. For example, NYCHA rules prior to 2002 allowed gardeners to plant only annuals and inhibited the development of long-term stewardship over sites. Gardens had to be mowed back each season and were not considered permanent. However, garden rules were revised in order to allow perennial plantings, enhance a sense of ownership, and promote sustainable gardening practices (Bennaton 2009, 236). The program has worked to support existing garden sites by helping to build raised beds, installing rainwater-harvesting systems at selected sites, and working to strengthen connections between resident gardens and community centers on NYCHA grounds. In 2015, the *NextGeneration NYCHA* ten-year strategic plan was released; it emphasized a "transition from direct service provision to a partnership-based model," meaning that cuts to the Garden and Greening program's staff and consultants were made, with an emphasis on working with outside partners, such as the nonprofit Green City Force, to develop urban agriculture sites (NYCHA 2015, 93). Despite this reorganization, residents are still allowed to garden on NYCHA grounds, and the citywide NYCHA garden competition continues.

Urban Agriculture's New Wave in NYC: Mid-2000s to Present

National and local media, celebrity engagement, and new funding streams were all indicators of rising attention to urban agriculture and local food during the 2000s. At the national scale, this included First Lady Michelle Obama's White House organic garden that was created in the spring of 2009; funding from national foundations, such as programs from Robert Wood Johnson focused on healthy eating and walkable communities as a response to the obesity and diabetes epidemics; and the highly popular food writing by author Michael Pollan and others. Celebrity chefs also played roles as public figures, advocates, and donors to urban agriculture programs—including Alice Waters of Chez Panisse, Dan Barber of Blue Hill, Mario Batali, and Rachael Ray (and her Yum-o! foundation). Local funders of note included the Doris Duke Charitable Foundation, which—in a shift from a previously rural environmental focus—gave seed money to several citywide projects related to urban agriculture starting in 2010. There is the potential for future funding to emerge from the Community Food Funders affinity group, which formed in 2011. These funders both stimulate more engagement in farming and are reflections of already-occurring changes in the social context—operating in a positive feedback loop.

The economic recession of 2008 influenced participation in urban agriculture in multiple, sometimes conflicting ways. Many respondents felt that the downturn attracted new constituencies of unemployed and underemployed people (or simply people looking to economize their food budgets) into gardening, canning, and food production as leisure, self-sufficiency, and cottage industry practices. In turn, whole industries have developed in support of these do-it-yourself (DIY) and locavore practices (such as stores and websites serving urban farmers and gardeners). This economic downturn may also have attracted young people into entrepreneurial urban agriculture endeavors, with individuals and firms selling produce and added-value products in farmers markets, through community supported agriculture (CSA), to restaurants, and at alternative markets like flea markets, pop-up events, and food trucks. Structurally, the slowed economy (at least temporarily) decreased development pressure in cities, meaning that more vacant lots were available for farming and gardening. However, interviewees were careful to note that the development pressures in New York City are quite different from those in, say, Cleveland or Detroit. At the same time, the recession placed additional pressure on formal nonprofits as they competed for scarce resources while attempting to fundraise for farming, environmental education, animal husbandry, and horticultural programs, many of which were facing increased participant demand. So, too, did municipal programs face budget cuts and staff reductions in response to fiscal belt-tightening at city, state, and federal levels.

The urban agriculture movement is far from conflict free. Axes of social difference exist in terms of race, class, gender, ethnicity, and cultural background—an issue that has been brought to the fore by food justice activists who focus explicitly on structural inequality along racial and class lines (BUG 2012; Romer 2012). The size and diversity of New York City's environment adds to the complexity of trying to build coalitions or make physical changes to that landscape. For example, one urban agriculture program leader noted that he can get to Connecticut as quickly as he can get across the city to visit other Brooklyn-based groups, so logistics can present challenges to collaboration. Huge variation exists in the conditions faced by growers working on different physical site types (e.g., rooftops, vacant lots, within parks). And organizational differences between grassroots, volunteer-led groups, and professionalized, formal nonprofits can create both exciting opportunities for collaboration as well as challenging obstacles to mutual understanding. Organizations, like people, differ widely in their values, expectations, work styles, communication patterns, skills, and needs. With nonprofit organizations working hard to keep their individual programs afloat, it makes coalitions like the NYCCGC, the Food Systems Network NYC (FSNNYC), or the Brooklyn Food Coalition (BFC) and collaborative efforts around producing documents like *FoodNYC, Food in the Public Interest,* and *FoodWorks* quite noteworthy.

Community gardeners and long-time garden activists vary in their approach toward the new wave of urban agriculturalists: they can see themselves as members of, in alliance with, or in competition with this movement. One circulating narrative is that "old-timers" and "newcomers" differ by generation, race, and class. Many low-income people (often African American or Hispanic) endured decades-long disinvestment, crime, and violence in their communities and used gardening as a neighborhood stabilization strategy, in addition to a food-provisioning strategy. Now, New York City is booming economically in comparison to the 1970s and 1980s when many gardens were founded. Consequently, there is a new demographic of white, affluent, educated people engaging in urban agriculture—often via the context of professionalized nonprofits or entrepreneurial ventures. One interviewee quite simply called them "the hipsters" (respondent 39). A bureaucrat reflected on these divisions:

> It feels a little bit like a splintered movement to me right now. I think there are the old-school community gardeners who tend to be New Yorkers who have been here a long time. They remember when their neighborhoods were devastated and they took back the land. . . . I don't mean to be too negative, but I think there is definitely a difference between the hipster, locavore movement that is starting now with younger people, and the old-world, community gardeners and how those two end up interfacing and interacting, is yet to be determined. (respondent 50)

The new wave of engagement has been met with increased attention by the media. Stories focus on the use of technologies (green roofs, aquaponics, greenhouses), charismatic individuals (ex-financiers, self-starting entrepreneurs, urbanites fleeing to rural farm life), and quirky farm stories that feel out of place in an urban context (keeping chickens and bees) (see, for example, Severson 2008; Salkin 2008; Cardwell 2010; Ryzik 2009; Stein 2010; Wells 2010). Some interviewees noted a disparity in media coverage and attention between longtime community gardeners and next-generation urban agriculturalists: "There's really a big disconnect between some of the white middle class practitioners and African-American, sometimes poorer [practitioners], sometimes just based on race distinctions that exist in terms of access to funding, access to city land, and other resources" (respondent 12). Others disagreed and argued that the increased attention is good for the movement on the broadest level, or felt that there is no reason for conflict among the various constituencies engaged in the movement. It is clear that urban agriculture is not comprised of a single, narrow constituency, but rather consists of diverse sets of people interested in managing urban land and changing the food system and their role in it from very different vantage points.

Storylines and Practices of Urban Agriculture and Local Food

New narratives and practices are developing around urban agriculture and local food systems. At the same time, old narratives are revisited and re-presented for twenty-first-century audiences. I identified several distinct, if overlapping, focal areas of the local food movement in New York City that have left an imprint on recent urban agriculture efforts. It is crucial to understand these narratives in order to contextualize the policy efforts that I describe in chapter 6—which arguments get taken up, which get bracketed out, who wins, and who loses. These themes include *locavorism and alternative food networks; food access, food security, and food justice; lack of data and lack of space in the developed city;* and *environmental education, community empowerment, and strengthening of local economies*—each of which will be examined in turn. These storylines and practices are woven together primarily via a civic-led movement of individuals, community-based groups, formal nonprofits, and foundations along with corporate funders, elite and public sector allies, and the media. Some members of the movement focus on individual behaviors (e.g., of consumers); others focus on developing new markets, institutions, and policies; and still others focus on structural inequalities and the need to redress those inequalities.

Locavorism and Alternative Food Networks

Locavorism—an emphasis on eating locally or regionally—is one of the main narrative themes of the contemporary food movement. Although one of the most hyperlocal forms of food *production* for New York City is urban agriculture, locavorism focuses primarily on the *consumption* of local—or regional and statewide—foods by New York City residents. For example, a trope that gets used in policy and journalism arenas is the claim that New York City residents should not be eating apples (or drinking apple juice) from China or New Zealand, given that New York State is the number two producer of apples in this country. An advocacy group's policy memo in response to PlaNYC, the *Food in the Public Interest* plan, and the *FoodWorks* plan all reference this concept. Interviewees, too, repeated the trope, saying, "So we can grow food in the city that can be, by far, healthier than an apple shipped from New Zealand" (respondent 45). This vignette is evoked repeatedly, banking on the intuitive sense that something is wrong or unhealthy with eating globally, particularly when food is assessed for its "food miles"—or how far it travels from farm to plate. The concept of food miles was, in part, developed by health scholar Joan Gussow (1999, 2006; see also Gussow and Clancy 1986). It was further popularized by nutrition scholar Marion Nestle, author of numerous books on how we eat, including *Safe Food: The Politics of Food Safety* (2010) and *What to Eat* (2005).

Locavores see eating as a political act, in which one can "vote with one's meals." Eating locally is seen as a way in which consumers and restaurateurs can support regional farmers and as a way of producing food that differs from the global, corporate agribusiness model that has grown dominant in the second half of the twentieth century (Lyson 2007). It is part of a broader change in food culture, what one respondent called "a general Zeitgeist of food":

> But I think the general trend is more heartening. You see more young people in the [farmers] market. You see hipsters in the market. . . . The worst thing to ever happen to farmers was Tang . . . because we had this whole culture in the United States where people came from farms. Their parents worked on a farm. And even if they moved to the city, they had a relationship with the farm growing up. And then we had this whole revolution of food where we're going to package everything and dehydrate it and send it to the moon. And then everyone started thinking that you could break your food down into, you know, five nutrients and call it food. So, Tang being the perfect example of an orange flavored beverage with all the vitamins and nutrients that an astronaut needs. So, we have this group of people—the boomers or the end of the boomers—who were without a food culture. And now we're slowly rebuilding that. (respondent 34)

Attention to the health or toxicity of food has also increased, with discussions of genetically modified, conventional, and organic foods being heightened by food scares and contamination incidences. Individuals have agency as consumers who seek to "know where their food comes from"—another common rallying cry of the movement. This view has been advanced by public figures like Pollan, Waters, and Barber. However, some advocates counter the prominence of Pollan in the public discourse, noting, "He's verbalizing the movement, he's not creating a movement" (respondent 34). The Slow Food movement has also advanced the locavore view transnationally; it has particularly strong chapters in Europe, as it originated in Italy (Slow Food USA 2012). Critiques of the industrialized and globalized food system have been further popularized through documentaries like *Fresh* and *Food Inc.*

A cohort of farm-to-table, seasonal, and slow cuisine restaurants, chefs, and foodstuffs has matured in New York City and nationwide during the 2000s. This shift is particularly notable when contrasted with the haute cuisine of the 1990s that focused on global, exotic foods. Numerous interviewees felt that this culture of food has reached a tipping point, as hyperattention to food origins by restaurant-goers is parodied on television shows like *Portlandia*. Other chefs work hard to disabuse the public of the notion that local, seasonal, and fresh food is only for stereotypical, white "hipsters." For example, chef and activist Bryant Terry writes about African American traditions of cooking and eating in books like *Vegan Soul Kitchen* and was one of the keynote speakers at the 2012 Brooklyn Food Conference, where he acknowledged the work of activists of color and the need for a social movement around food justice. Finally, local restaurants in New York City participate in the urban agriculture movement by hosting fundraising dinners (such as ici and Good Fork for Added Value) or donating materials (such as used coffee chaff from Café Grumpy) as bedding for chickens at Eagle Street Rooftop Farm.

Small-scale, young, and new farmers in rural areas proximate to urban centers are important participants in local and regional food systems and conversations around sustainable agriculture. Although there is an overall decline in the number of farms nationwide—from more than six million farmers in 1910 to just over two million farmers in 2007—due to farm consolidation and loss of farmland to development, there is also a countertrend of young farmers developing small farms, often in close proximity to urban areas and their robust consumer markets (Shute 2011). In New York State, the average age of principal farm operators is 56.2 years old (Cohen 2011). The national trends are striking:

> For each farmer under 35 there are 6 over 65 and the average age of farmers is 57. It is estimated that between now and the year 2030, half a million (one-quarter) of American farmers will retire. . . . As public awareness about food and farming has grown in recent years, a new opportunity for reversing the trend of farm loss has arisen. The National Young Farmers Coalition is witnessing an increasing number of young people from non-farm backgrounds who are pursuing or considering careers in agriculture, as well as a higher level of interest among farm youth in staying on, or returning to, the family farm. (Shute 2011, 9–10)

The Greenhorns is both a documentary film about this phenomenon and a loose collaborative formed in 2007 to use media, social networking, and grassroots organizing in support of young farmers (Greenhorns 2012). The National Young Farmers Coalition helps serve that same constituency, organizing and advocating for policies and practices that support young farmers and rural livelihoods (NYFC 2012). Small farms face common struggles: access to capital, land, and credit; costs of inputs; logistics of transportation and selling; and the need to price products in order to make a profit (GrowNYC 2011, 15–17; Shute 2011). The margins of profitability are so small that there is sometimes tension between their needs as emerging entrepreneurs and their desires to sell affordable food in underserved urban areas; or between their need to grow in order to be profitable and their desire to remain small scale. Some state and federal programs have developed to support these newcomers, including programs via USDA's National Institute of Food and Agriculture. Funding and technical assistance are available through the Beginning Farmer and Rancher Development Grants and Community Food Projects (USDA 2012).

Farmers markets and CSAs provide mechanisms to develop relationships between rural farmers and urban consumers. In New York City, the largest farmers market system is the Greenmarket, which is operated by the nonprofit GrowNYC. It began in 1976 with one market in Manhattan's Union Square Park and has since expanded to more than fifty markets citywide, engaging more than 230 farmers (GrowNYC 2010, 2011). However, Greenmarket does not control all of the farmers markets within New York City; there are twenty-eight different institutions operating fifty-eight distinct community-based farmers markets citywide that are outside of the Greenmarket system (Stringer 2011).[7] Concurrently, CSAs are on the rise locally and nationally, with more than 137 CSAs in New York City as of 2016. This model allows members to buy shares of a farmer's harvest, entitling them to fresh produce (or cheeses, meats, breads, honey, eggs, milk, flowers, and added-value products) as it is available throughout the growing season. The nonprofit Just Food was instrumental in growing the CSA presence in New York City; prior to that organization's founding there was just one CSA citywide. One respondent noted the way in which Just Food's work supports both urban and rural constituencies, "The CSA program is really designed to serve both the community and the farm. A high-income neighborhood serves the farm very well, so if it's supporting a sustainable family farm within the region—that works with our mission. The urban agriculture work that we do tends to be very focused on community gardens . . . lower-income neighborhoods" (respondent 60).

Despite all the opportunities provided by the size of the New York City consumer market and the number of farmers markets, rural producers still struggle with profitability in part because of the logistics of transporting and selling within the city. There is sometimes a mismatch between the producers' need to sell a certain volume and community efforts to bring markets to previously underserved areas. As such, there are long waiting lists amongst producers to sell at the most profitable farmers markets. One solution is to become large enough to sell at multiple markets, but this runs counter to some producers' interest in maintaining small-scale agriculture. Moreover, New York City does not currently have a permanent wholesale market for small regional farms that would allow growers to sell their goods to urban commercial markets more efficiently and at greater scales year round. GrowNYC has helped to create the existing wholesale farmers market presence at the Hunts Point Food Terminal in the Bronx. That market currently involves approximately a half dozen producers and is located in the parking lot of the Hunts Point market, without necessary infrastructure and services to allow the market to operate at a larger scale (NYC Council 2010, 21). Discussions around creating a permanent wholesale market and food hub are increasing among policymakers and advocates alike (Severson 2008; Stringer 2009, 2010; NYC Council 2010; Navarro 2012).

Beyond market relationships, some policy alliances are beginning to form along upstate-downstate (or rural-urban) lines. For example, in 2011, meetings and rallies were organized for New York City residents to get involved in Farm Bill advocacy through the NYC Food and Farm Bill Working Group. New York City–based activists worked with the American Farmland Trust on upstate land preservation issues and hosted Senator Kirsten Gillibrand for a series of public meetings. And in 2009 and 2010, upstate growers came to the city to attend conferences on food policy and food systems (described in chapter 6). Programs in support of New York's regional food systems and economic development (including food processing, industrial retention, and support for regional grain production and bakeries) have been developed and supported by the New York State Department of Agriculture and Markets.

As new programs and funding streams have developed around local and regional food systems, so, too, has the scientific, policy, and activist debate around locavorism. Indeed, even the "apple trope" discussed earlier has been exposed to questioning. A policymaker said, "The question is, right: we have a million kids in New York City public schools. Does it *matter* if they eat an apple from New Zealand or if they eat apples from upstate New York?" (respondent 52). Critiques of locavorism include the notion of food miles as an inappropriate measure of environmental impact; the limits to locavorism, particularly in less hospitable growing climates; and the efficiencies of global trade established by neoclassical economics (McWilliams 2009; Budiansky 2010; Desrocher and Shimizu 2012; Morgan and Sonnino 2010, 212). There are efforts to quantify and understand the benefits and tradeoffs of local versus global food systems in terms of energy use, carbon emissions, water use, and climate change (see, e.g., Canals et al. 2007; Weber and Matthews 2008). How to weigh these tradeoffs remains a classic conundrum of sustainability planning. Another debate centers on the environmental gains and efficiencies of densely populated cities versus the importance of having open space and agricultural space nearby (Owen 2004, 2009; Glaeser and Kahn 2008; Bettencourt and West 2011). This question hinges on what sort of a city is desirable and how we define sustainability.

Some scholars and activists question localism from a social justice perspective. Geographers critique the a priori privileging of the local scale in addressing food system injustices; others critique a "defensive localism" that alienates outsiders; and others call for a "reflexive localism" that forges alliances that attend to social justice (Hinrichs 2003; DuPuis and Goodman 2005; Born and Purcell 2006). There is also concern that locavorism is an elitist practice, or associated with an elite stigma (McWilliams 2009). Locavorism and elitism became further linked through the practices of high-end retailers like Whole Foods, which sell fresh, organic, seasonal fare at premium prices—jokingly mocked as "Whole Paycheck" by activists (Gottleib 2010, 3; but see Singer and Mason 2006, 5). Others argue that the emphasis on local foods misses greater structural inequalities and the machinations of capitalism that drive food injustice, and question the premise of "voting with one's meals" (Guthman 2011).

Finally, as notions of sustainable urbanism become more mainstreamed, there is potential danger in corporate cooptation or "greenwashing." Corporations have sponsored community gardens as a way to associate their brand with the urban greening movement.[8] In addition, corporations have tried to harness enthusiasm for eating locally and supporting family farms through marketing campaigns focused on the source of their ingredients. These corporate entities are seeking ways to align themselves with a hip movement that can enhance their image as "green." One sustainable agriculture advocate discussed Lay's potato chips and the loss of meaning in the term "local food": "You have these advertisements where it's like, 'Oh, and our family farmers are local, we're in Texas. We grow potatoes in Texas; we're in Texas so eat your Texan Lay's potato chips.' But the fact is that that language is actually being picked up and becoming part of the lexicon: 'Buy local. Buy local'" (respondent 33). Various labels of products as "local," "organic," or "natural" can lead to numerous, complex choices for consumers. One respondent noted that the term CSA is being utilized by some groups to refer to arrangements that are quite different from the original intent of the CSA. Labeling schemes can serve as both feedback mechanisms for the consumer and marketing devices. Overall, the use of the language of "local food" by activists, policymakers, the general public, and corporate entities is becoming increasingly loaded with mixed and even conflicting meanings.

Food Access, Food Security, and Food Justice

Interest in access to healthy food is not limited to upper- and middle-class "foodies" with disposable income for premium produce. Advocates have organized around the issues of food security and access for low-income, underserved populations. This issue is sometimes framed around the twin crises of obesity/diabetes and hunger, which some have dubbed "the Bronx Paradox" (Dolnick 2010). National leadership in combatting obesity and diet-related diseases has come from foundations, particularly the Robert Wood Johnson Foundation and the Kellogg Foundation, as well as from First Lady Michelle Obama via her "Let's Move" campaign and her White House garden. Kellogg has supported the development of local Food and Fitness Partnerships, including one in Brooklyn. In terms of hunger and food security issues, local nonprofits like City Harvest and the Food Bank for New York City work to feed hungry people and connect them to resources, while networks like NYC Coalition Against Hunger,

WhyHunger, and the Hunger Action Network of New York State lead policy conversations and provide support to antihunger groups in the movement.

In New York City, municipal leadership on these issues came from the Department of Health and Mental Hygiene (DOHMH) and its District Public Health Offices; former city council speaker Christine Quinn's office; former Manhattan borough president Scott Stringer; and former mayor Bloomberg via the food policy coordinator (FPC) (see chapter 6). Indeed, Stringer's *Food in the Public Interest* report begins with a call to action framed through the obesity crisis:

> Problems associated with obesity and overweight have reached epidemic proportions in the United States. . . . New York City is outpacing the nation in obesity and its related health issues. Both obesity and diabetes rates rose by 17 percent between 2002 and 2004 among city residents. It is estimated that New Yorkers gained more than 10 million pounds collectively during this same period. A rise in the risk of heart disease, hypertension, depression, type II diabetes, among other health problems, often accompanies a rise in obesity and overweight. Residents of low-income neighborhoods and Black and Latino adults are disproportionately affected, thus overburdened by the related health, social, and economic problems. (Stringer 2009, 4)

Many of the solutions being offered by city officials focus either on individual consumption patterns or on incentivizing grocery stores (and mobile fresh fruit vendors) to locate in low-income neighborhoods with poor healthy food access. In addition, framing gardens and urban farms within the context of hunger and obesity crises increased attention to these efforts, a respondent noted: "But what changed all of that, from the Deputy Commissioners all the way up to the Mayor, was the sudden huge emphasis on food and childhood obesity. Once it became all about food, everybody wanted a piece of that. . . . And when community gardens stopped being looked at as this kind of not that nice, bohemian, self-managed pocket park program and as a food program, all of a sudden we're important. . . . In terms of how much support we got, it was like night and day" (respondent 6). Overall, the crises of obesity and diabetes have lent a new urgency to longstanding issues of food *access*, hunger, and healthy eating and are beginning to be woven into discussions around urban food *production*. Starting in 2012, a mayoral Obesity Task Force launched a program to expand gardens and farms in neighborhoods with high incidences of diet-related diseases and to support additional school gardens citywide.[9]

Unique collaborations between foundations, nonprofits, city agencies, and elected officials have developed to address the people and areas with poor access to healthy, affordable, and fresh foods. These areas are sometimes referred to as "food deserts"—although this is a contested term by academics and activists (see, e.g., Huber 2011; Fields 2013; Shannon 2013). GrowNYC has coordinated a number of efforts to address hunger and food inequality via the Greenmarkets. Beginning in 2005, they initiated one of the largest efforts in the country to accept Electronic Benefit Transfer cards (food stamp debit cards) at farmers markets. This program has been acknowledged as a best practice and is now emulated in several cities. In 2011, more than $600,000 in Supplemental Nutrition and Assistance Program (SNAP) funds—food stamps—was spent at Greenmarkets (Van Ooyen 2012), though several billion dollars in SNAP funds are received and spent in New York City each year (FSNNYC 2010). The Greenmarkets also accept HealthBucks, which are $2 vouchers for fruits and vegetables redeemable at farmers markets; for every $5 spent on fresh produce at farmers markets using EBT, participants receive a HealthBuck. This program was created via the DOHMH and distributed through citywide and neighborhood-based social service providers (NYC DOHMH 2012). After the Hunts Point Food Terminal, Greenmarket is the second-largest donor of food citywide to City Harvest food pantries (respondent 34).

While everyone has the right to healthy and fresh food, particular attention has been focused on children in New York City, for multiple reasons. First, childhood obesity has reached epidemic proportions, with dire lifelong consequences for individuals (Stringer 2009, 2010; NYC Council 2010; Romer 2012). The 2007–2008 National Health and Nutrition Examination Survey of the Center for Disease Control and Prevention (CDC) found that 16.9 percent of the U.S. population between the ages of two and nineteen is obese (Ogden and Carroll 2010). Second, the problem is particularly acute for low-income children, with one in seven low-income preschool-aged children being obese and nearly one-third being overweight or obese (CDC 2009). In terms of variation by race and ethnicity, the CDC notes that in 2009 "American Indian or Alaska Native children had the highest prevalence of obesity (20.7%), followed by Hispanic (17.9%), non-Hispanic white (12.3%), non-Hispanic black (11.9%), and Asian/Pacific Islander (11.9%)" (CDC 2009). The focus on urban, low-income, and minority children as vulnerable populations in need of care has an abiding tradition in activism and social services in the United States since at least the Progressive Era in the nineteenth century. Third, contemporary youth organizations have organized around the food justice theme at national conferences, such as the Rooted in Community Conference, and through local organizations, like Flip the Table Youth Food Council.

Changing school food procurement and preparation practices is one key lever that municipal government has in altering the food system. As such, it has been aggressively targeted by advocates and policymakers. The Community Food Security Coalition, one of the major national food security networks, has organized a National Farm to School Network to work on policy advocacy, the Farm Bill, and program development. Farm-to-school programs in New York City have the potential to make a large impact, both because of the buying power of the DOE and the opportunity to reach such a large group of young people, with about half the population of children in New York State living in the city. Indeed, the DOE serves 860,000 meals per day and is the second-largest institutional procurer of food in the nation after the U.S. military (Cohen 2011). The DOE SchoolFood program has worked to increase the amount of fresh fruits and vegetables available in public school lunches, and the Garden-to-School Café program focuses on serving fresh vegetables grown in school gardens as part of school lunches. Local legislation passed by the city council as part of the *FoodWorks* plan (described in chapter 6) required the DOE to report on the amount of regionally sourced foods, as one step in working to change procurement practices. Finally, the Grow to Learn effort created by GrowNYC and DOE supports the development of school gardens citywide, with 224 school gardens created and registered with the program as of 2012. There are countless school-based education programs focusing on gardening on school grounds, in nearby community gardens, on rooftops, and in greenhouses.

Food justice activists similarly seek to remedy access and insecurity issues but adopt a more radical focus on shifting the balance of power within the food system. These activists position access to healthy, fresh, and affordable food as a right—a right that has been compromised because of social inequalities along axes of class and race. As such, they critique urban agriculture efforts that do not address inequality issues as potentially counterproductive or confusing to the public: "I do think there's been a growing number of groups that grow fresh food in the city but don't necessarily do food access work. And I think that . . . it's easy for people to confuse that. . . . There's sometimes groups that are very happy to be like, 'Yeah, we're in . . . this neighborhood that has no fresh food. And we grow fresh food.' And they don't ever say, 'But our food doesn't go to the neighborhood.' But that's the reality" (respondent 43). Similar to the way in which environmental justice reinvigorated environmentalism and brought new constituencies to the table, so, too, is food justice creating the next wave of the environmental movement, with attention to these issues rising during the 1990s and 2000s. Bryant Terry framed this engagement as a continuation of the tradition of African American leadership and organizing around food, citing how the Black Panthers provided free breakfasts in low-income communities of color in the 1960s and how the Nation of Islam emphasizes healthy diets (Terry 2012).

Food justice activists have raised the issues of process, inclusion, and representation in food policy arenas. For example, the founders of the Black Urban Growers (BUG) coalition seek to ensure that people of color have a seat at the table in food policy decision-making (BUG 2012). Others draw attention to the racial composition of the staff of professionalized nonprofits in urban greening, including one activist's own organization: "I think because we were at the front end [of the urban agriculture movement] and we were in a gentrifying neighborhood and because we—like most of the institutions that are involved in paid work in agriculture—are white . . . we have been the focus of some of the conversation around racial equity in urban agriculture. . . . Some of that has to do with the fact that we don't have staff who are from [our neighborhood]. But nobody does in any of the paid organizations" (respondent 23). Some organizations have worked to address this disparity by shifting authority to leaders in the communities they serve:

> Our Trainers of Trainers Program recognizes the knowledge and expertise of urban farmers and community gardeners in NYC—particularly people in the communities that might be considered food deserts. . . . It's actually even more powerful to have people leading workshops on growing food and food justice if they are from your own community and they lived a similar experience that the folks in the community lived, versus having staff of an organization, which is usually young and white and professional, or at least middle class. (respondent 21)

This "train the trainer" model emphasizes community-based leadership and pays trainers a $100 stipend per workshop delivered. Members of the food movement noted the challenge of scaling up organizations that are truly grassroots and community-based, wrestling with how one retains community control and democratized decision-making while also trying to expand the reach of programs to serve more New Yorkers. In sum, activists draw attention to the question of "who speaks for whom" in urban agriculture programs and policymaking.

Because of the movement's focus on social justice, alliances have been fostered with the global food sovereignty, labor, and antiracist movements. At a public panel, Ray Figueroa, a food justice activist affiliated with NYCCGC, offered the following commentary: "Food justice is the servant of food sovereignty, which is the utopian ideal of sustainable subsistence in harmony with nature" (BUG 2012). Another activist elaborated on the multiscalar organizing around food, health, and land: "Oh, definitely, it's a social movement. I tell people this is the next civil rights movement around food: food sovereignty. Because it's not happening just in New York; it's happening all over the country. It's happening globally. People are really looking at where their food comes from. . . . They understand . . . how land and food and health and wealth and economics and culture all interlock" (respondent 29). The global reach of the movement is evidenced by two of the keynote speakers at the 2012 Brooklyn Food Conference: renowned Indian activist and scholar Vandana Shiva (by video address) and Luis Benitez, a migrant farmworker who organized for tomato harvesters' rights in Florida. Shiva and Benitez were seen as working in solidarity with Brooklyn activists. They discussed improving worker conditions and providing a living wage as the surest routes to alleviating food insecurity. Attendees at the conferences cheered in response to speakers' references to living wages—as fast food retail workers in New York City worked for rights and protections just like the migrant workers in Florida. Furthermore, "Dismantling Racism" trainings have been offered via the Community Food Security Coalition to activists across the country. These trainings seek to grow personal and organizational awareness of racism and white privilege. In addition, the nationally prominent urban agriculture organization, Growing Power, created a program entitled "Growing Food and Justice for All" focusing explicitly on strengthening the network of groups involved in food justice and anti-racist efforts (GFJI 2012). Growing Food and Justice for All's first national conference was held in 2008 and included approximately thirty delegates from New York City organizations, including many that were sponsored to travel by the organization Just Food (Heehs 2008).

Starting in September 2011, food justice groups were engaged in the Occupy movement. A "Farmers' March" was held on December 4, 2011, under various mantles of "Occupy Food" or "Occupy the Food System." Food justice was made an official working group under the Occupy Wall Street General Assembly. The closing plenary of the 2012 Brooklyn Food Conference by BFC founder Nancy Romer made a call for just these sorts of alliances:

> Let's look at who our real allies are. . . . There's a growing movement of people who have had enough, people who know that their lives are not improving, people who are losing ground economically, socially, in their families and communities. Some call it the 99%, others call it the weak vote, but we must see our interests as allied, and we must participate in the movements to protect our people and our planet. . . . Let's join the environmental movement with the anti-racist movement, with the anti-war movement, with the Occupy movement, with the labor movement, with the women's movement, with the LGBTQ movement. (Romer 2012)

What's Missing: Lack of Space/Lack of Data

For all its regional, national, and global reach via social movement networks, the local food and urban agriculture movement in New York City must be understood in its historically path-dependent and spatially specific context. Local decision-makers consistently voiced the sense that the city is already so developed that there is a lack of space available for urban agriculture. Public officials noted that "this is not Detroit" and that space is at a premium in New York City (respondent 34). One bureaucrat elaborated, "Whatever values one ascribes to urban agriculture, it has nothing to do with large pieces of land. Not in New York. Because the land isn't available. In Detroit now, yes. . . . And in the cities that are truly in the same condition that New York was in the 1970's when large amounts of property . . . were being abandoned, not maintained. So you go to Youngstown, Ohio. You go to Flint, Michigan. You go to Detroit. You go to Cleveland. There's an opportunity in those places that we had in the seventies" (respondent 41). A 2011 study estimated that there were approximately five thousand acres citywide of vacant land potentially suitable for urban agriculture (Ackerman 2011).

Space constraints have led to experimentation with alternative growing sites, including rooftops, backyards, and temporary spaces. A rooftop farmer offered stark observations of the economics of their operation: "One thing that is very different *now* [as compared to] farming in the city 150 years ago is there's no space. And the space that exists is enormously more valuable as real estate . . . than it is for food. The only reason I can compete with the space around me is that I'm on a roof. If I was doing a quarter-acre of production on the ground it . . . could be turned into apartments that would be worth $3 million. There's no way I'd be able to farm" (respondent 44). Rooftop farms include both intensive hydroponics and greenhouse agriculture (Gotham Greens, BrightFarms) as well as growing on green roof media and lightweight soils (Brooklyn Grange, Eagle Street Rooftop Farms). Although these firms are few in number, they have received a significant amount of media attention and coverage (see, e.g., Cardwell 2010; Stein 2010). Images of the rooftop farm set against the Manhattan skyline circulate and inspire imagination in news features, websites, blogs, student design presentations, research reports, and policy documents (see figure 5.3). Rooftop farms seem to offer one way around the conundrum of lack of space, not competing with but enhancing the value of the residential and commercial buildings below.

Gardening and animal husbandry is expanding in small sites across the city. While backyard gardening is nothing new, gathering these disparate sites together and giving access to gardeners without sites is an innovation. This is the approach of Brooklyn-based nonprofit BK Farmyards, which developed a creative solution to an old problem. A recent report quoted Nevin Cohen's estimate that "New York City has 52,000 acres of backyard space that collectively could provide vegetables for 700,000 people" (Stringer 2010, 11). The number of people keeping chickens in their own private backyards as well as in community gardens has also increased, facilitated in part by Just Food's City Chicken program. Beekeeping

FIGURE 5.3 Brooklyn Grange Rooftop Farm. Photo by author.

has also proliferated since activists worked with the DOHMH to change a local law that prohibited beekeeping within city limits in 2009. Several other nontraditional innovations designed for growing in compact spaces include vertical farming, window farms, hydroponic growing, aquaculture, and grow boxes (or "Earth Boxes"). These approaches range from low-cost, low-tech solutions that are implementable now to more high-end, expensive, or futurist visions.

Because of the history of crisis and development pressure, many community gardeners distrust temporary sites. That attitude has begun to shift, however, as some gardeners and farmers are willing to access sites without strong guarantees of tenure or protection, with the understanding that they may only have their farms for one or several growing seasons. In some cases, when housing development is on a slow timeline, HPD authorizes the temporary use of sites as gardens. For example, A Small Green Patch is a temporary community agriculture site in Park Slope, Brooklyn, that used large fabric container plantings to grow food, with the understanding that at some point those planters would have to move when the site is developed. In another case, the Cypress Hills Local Development Corporation (LDC)—a nonprofit involved in community development that includes affordable housing, social services, and education programs—approached HPD about temporarily creating a garden on one of their own housing sites that was,

FIGURE 5.4 El Jardin del Pueblo, a raised-bed garden organized by Cypress Hills LDC on land that temporarily could not affordably be developed into housing. Photo by author.

at the time, too small to develop affordably into housing. The site became a flourishing garden named El Jardin del Pueblo with raised beds, chickens, a rainwater-harvesting system, and an active youth and adult membership (see figure 5.4). As a result, there had been some preliminary discussion about the possibility of transferring the site permanently to DPR. However, as of 2016, the site has been identified again as a potential housing development site under the de Blasio administration. This site demonstrates the tenuous nature of growing on HPD land.

Perceptions of both a lack of space for agriculture and a lack of data on urban agricultural systems in New York City have led researchers and activists to initiate studies that gather and analyze data about open space, gardens, and farms. This work was both self-motivated by academic and civic scientists as well as emerging in response to the demands of policymakers. The pressure for measuring urban agriculture's potential and its impacts is quite great, despite the acknowledged complexity of the task. Two notable studies include the Columbia University Urban Design Lab's "Report on the Potential for Urban Agriculture" and the Five Borough Farm, a joint project of the Design Trust for New York City and

Added Value, working with professor Nevin Cohen. The Doris Duke Charitable Foundation funded both of these projects in 2010, and both sought to assess the feasibility for scaling up urban agriculture (Ackerman 2011; Cohen et al. 2012). On a larger geographic scale, the Columbia Urban Design Lab conducted a GIS-based study on the greater–New York foodshed (defined, variably, as the land within one hundred or two hundred miles of New York City) to assess its capacity to produce, process, and distribute different types of fruits, vegetables, and meat (Conrad et al. 2011). These studies were some of the first citywide efforts to take on the challenge of mapping and quantifying aspects of the regional and local food systems into forms that would be legible to policymakers.

Other projects use GIS and mapping techniques in a participatory manner to support community-based planning or to help organize the reclamation of vacant lots. In the past, using geographic data required an understanding of and access to software like ArcGIS; but public GIS projects like OASIS and 596 Acres make these data available to anyone with a web browser. OASIS began in the early 2000s as a place to share geographic data long before the municipality began making those datasets more readily available for download; and it is also a site where other data generators besides city agencies (such as nonprofits and researchers) can serve up their data. Working with GrowNYC, OASIS was the first website to map community gardens citywide across all land jurisdictions, recognizing the validity of gardens as a land use and making them visible in new ways. It is also the host of the U.S. Forest Service's citywide Stewardship Mapping and Assessment Project (STEW-MAP) data that maps the spatial turf and organizational focus of hundreds of civic environmental groups citywide.

Founded in 2011, 596 Acres made an easy-to-navigate online map of all publicly owned vacant land in Brooklyn (which, at the time, totaled 596 acres). The goal of 596 Acres is to raise awareness of vacant lots, facilitate organizing the sites for alternative uses, and create pressure on agencies that own these sites. While the city has a database of vacant properties, PLUTO, planners know it to be fraught with errors and frequently out of date. As in many cities, there is a need to field-verify whether the sites shown as vacant in the database are, indeed, vacant in real life. This group reinforces its online strategy with public signage that it places on vacant lots through volunteer efforts. In many cases, these sites are held by HPD, which intends to sell them to developers for housing development. While 596 Acres founder Paula Segal described this practice as "warehousing lots" (Segal 2012), a public official noted that many of these sites are often in various stages of the pipeline for development, as HPD often has programs for lots on five-, ten-, and twenty-year pipelines. Building on the diverse traditions of guerilla gardening, "Right to the City," and Occupy Wall Street organizing, 596 Acres had helped people convert ten previously vacant lots to productive use, as of 2013. The group then worked to expand its scope from Brooklyn to the entire city (596 Acres 2012). By 2014, 596 Acres was working with activists located across the country and the world, advising on satellite projects in New Orleans and Los Angeles and hosting a multiday conference called "Vacant Acres" at the New School to bring like-minded activists together on issues of land access and land tenure. Results from this conference were later shared via the open access online journal *Cities and the Environment.*

Civic science projects—such as Farming Concrete, Seeing Green, and FarmingUp—work to measure the impacts of urban agriculture through participatory approaches. Farming Concrete began as an effort to quantify the produce grown at community gardens, working collaboratively with gardeners to weigh and log their harvest. This group developed open access tools that it hopes to spread to other cities, offering a participatory way to inventory one aspect of the productivity of gardens (Farming Concrete 2011; Gittleman et al. 2012). Working in collaboration with Five Borough Farm, Farming Concrete expanded their scope to include other impacts of gardens, including individual wellbeing and social outcomes such as changes in attitudes, knowledge and skills, and mood. Seeing Green complements Farming Concrete by quantifying the environmental impacts of urban agriculture sites. Organized by two amateur scientists and funded, in part, via a Kickstarter campaign, Seeing Green worked with Added Value and the Eagle Street Rooftop Farm to measure and model storm water retention on these sites using data collected on site from a lysimeter and moisture meter—and then modeling these impacts citywide by using data from the Columbia University report on urban agriculture (Seeing Green 2013). FarmingUp focuses solely on rooftop agriculture sites and has assessed the nutritional content of produce grown on rooftops. Because rooftop agriculture is such a new practice, much remains to be understood about the long-term viability of the soil media and the health and character of the produce grown on these systems. Other civic science projects focus on gardens as sites that enhance biodiversity, such as the American Natural History Museum's bee-watchers program. Taken together, these efforts aim to improve understanding of the multiple impacts of urban agricultural sites.

Beyond Food: Economy, Education, and Empowerment

Building upon the notion of the lack of space in the city, another trope circulating in policy, media, and activist circles is the claim that "you can't feed New York City through urban agriculture." In part, this is a simple acknowledgment of the reality of a city with more than eight million residents living in approximately three hundred square miles, with 33.5 million tons of food/year coming into it (Barron et al. 2010).[10] The *Potential for Urban Agriculture* makes the limits of urban agriculture in New York City starkly clear: "Converting all of the potentially suitable vacant land in the city (conservatively estimated at 4,984 acres . . .) to agriculture with an average growing area of 70% of lot area could supply the produce needs of approximately 174,000 people with biointensive yields, which is a substantial number but obviously not sufficient to feed the entire city. While there is much more land potentially available than just vacant lots, it is clear that NYC should not nor cannot strive to be anywhere close to self-sufficient in supplying its fruit and vegetable needs (much less all foods)" (Ackerman 2011, 22). An urban agriculturalist acknowledged the importance of rural agriculture from an efficiency and productivity perspective, as compared to rooftop farming in New York City: "Do I really care about rooftop greenhouses? No. Or rooftop farming for that matter? Not at all. Get a tractor! . . . New York State could feed New York State. You can grow the grain. You can raise the pigs, the cows, the chickens. You can do it here. Why put a couple of million dollars into a rooftop when for a couple of million dollars you could own vast tracts of land and tractors and employ a bunch of people?" (respondent 23). A rooftop agriculturalist similarly recognized the economic limits of their operation but focused on the co-benefits of the space, saying:

> To me the value of creating an urban agricultural space is that you're growing food—that's important. But you're also growing people's self-awareness. You're growing environmental connections. You're creating a green space. But if you take that away, the cost of food is so cheap in this country, it doesn't make *any* sense to grow food in the city. . . . That space is so much better used in terms of finance almost any other way than like growing *carrots*, 'cause carrots are *so cheap*. . . . We make two dollars a square foot . . . $12,000 dollars a year on food. You could rent this roof for two months and you'd make that money. (respondent 44)

Other urban growers argued that the idea that food production only happens in one way and in one form is what led us to the current state of centralized, industrialized agriculture in the first place, and that multiple forms and spaces of food production are more sustainable and resilient.

Because the narrative circulates so prominently, urban gardeners and farmers sometimes distance themselves from the idea of "feeding New York City through urban agriculture." They signal their understanding of the complexity of the issues and seek not to be associated with unrealistic claims-making. One food advocate noted, "I don't think we're going to feed New York City with urban agriculture. But I think it's extraordinarily important for all sorts of reasons" (respondent 38). The *Potential for Urban Agriculture* report situates its calculations about food production in the context of all the other benefits of the practice (Ackerman 2011). One respondent framed the debate around the politics of upstate/downstate divisions and how urban agriculture might transcend that perceived division:

> I feel like there's . . . upstate versus downstate all the time. . . . Rural folks get defensive saying, "Well, you know, you can't grow all your food in the city anyway." . . . It does worry me when I hear people talk about urban agriculture as sort of growing all your food in the city, 'cause I strongly believe . . . that's not really the point. The point is . . . this educational aspect of urban agriculture. It's the organizing aspects of urban agriculture. It's the fact that you're building markets within the city to support rural development in a way. And having more people that understand where their food comes from can only do good to help support local food systems and to [help support] rural agriculture. . . . (respondent 33)

Another urban farmer talked about agriculture in the context of the city's ecosystem:

> I think the strongest thing urban agriculture can do for itself is define itself . . . yes, as food production, which is important. But also we're creating greenspace in neighborhoods. We're supporting community. We're creating a place where stormwater can drain, which is environmentally benefiting the city with a lot of concrete on the sidewalks. You do all of those things. You add all of those and you count them towards the dollar value of the space. And that's where it's logical. Right now there's a lot of focus on: "urban agriculture's going to solve diabetes, urban agriculture is going to solve asthma." And the reality is: you need an entirely encompassing plan—like parks. You need clean air. You need less trucking. You need a local food *system*. You need people going to Greenmarket and receiving WIC coupons to go shopping. You need the city supporting the planting of street trees. (respondent 44)

Overall, this claim that "you can't feed yourself through urban agriculture" is a reflection of the way in which contemporary discourses around urban agriculture have come to be so closely connected to food production and consumption—when, in fact, urban agriculture first emerged via community gardening, out of concerns over neighborhood health, safety, and livability, rather than as a reaction to problems with the global food systems per se. Current practitioners of urban agriculture are working to counter the narrative by focusing (again) on the social, environmental, economic, educational, and spiritual benefits of their work.

Many groups focus on urban farms as sites that can support empowerment and strengthen community economies. Historically, several of the programs in this area were catalyzed by the northeastern office of Heifer International. While Heifer works globally to strengthen family self-sufficiency through animal husbandry, their focus in New York City was on incubating viable community agriculture projects. They provided initial funding to Just Food, East New York (ENY) Farms!, Added Value, and the New Farmer Development Project of GrowNYC.[11] Employment, internships, and opportunities to sell produce at markets are crucial to community farm programs. Notable programs with a youth-empowerment focus include Added Value, the nonprofit that manages the Red Hook Community Farm, and ENY Farms!, a program of the United Community Center in East New York, both of which are located in Brooklyn (see figure 5.5). Added Value seeks to create short food chains and feedback loops to keep resources circulating in the Red Hook community. Thus, they engage partners in developing local farmers markets and CSAs and sell produce to local restaurants. Just as alternative food networks are developing between upstate farmers and urban consumers, so are urban farms forming their own, localized relationships between neighborhood residents (as consumers, volunteers, composters),

FIGURE 5.5 Youth working at ENY Farms! in East New York, Brooklyn. Photo by author.

program participants (staff, interns, volunteers), and area restaurants (purchasers of produce, providers of food scraps for compost).

Small business entrepreneurs—including both for-profit urban farms and producers of added-value urban agricultural products—participate in developing local economies. Farm entrepreneurs often see themselves as having a mission to support living wages for agricultural employees, as well as a broader educational mission. A rooftop farmer said, "Our goal is not to supplement or supplant all of the rural farmers. . . . The goal [is not] how can we make money off of our food system? Yes, we should be paid a fair wage and nobody should be slaves in the food system. But making big profits when there's human health and the environment at stake . . . should not be your primary focus. . . . The goal is not to feed New York City. The goal is to educate New York City and green New York City and feed some of us" (respondent 54). Indeed, the same respondent that critiqued rooftop farming in comparison to upstate farming noted urban agriculture's importance as a local job generator:

> I think [rooftop farms] are interesting. I think they have great pedagogical value, I think particularly for high-end food, which New York City is awash in. It makes sense. I mean where else in the country can you buy a pint of pickles for nine dollars right now? You can't anywhere else but in Brooklyn get like 30 different varieties. And, that's great. . . . If people can make pickles for nine dollars and other people can pay for it and that creates small micro entrepreneurship opportunities for lots of people, that's totally cool. (respondent 23)

There is some fluidity and hybridity between private and civic sectors around urban agriculture, particularly in the case of programs that emphasize the "triple bottom line" of economic, ecological, and environmental sustainability. For-profit farms sometimes incubate affiliated nonprofits in order to be eligible for foundation grants, as was the case with Brooklyn Grange and the educational nonprofit City Growers. So, too, is Growing Chefs, an educational nonprofit affiliated with Eagle Street Rooftop Farm. Some nonprofits provide training to potential future entrepreneurs. For example, the New Farmer Development Program of GrowNYC offers training and technical assistance to new immigrant farmers in New York City who are interested in growing at larger volumes in order to sell at Greenmarket. Furthermore, community gardeners in New York City are permitted to sell agricultural products at farmers markets, provided that the revenues are reinvested into gardens. While permitting a new community market or Greenmarket is somewhat complex, any community garden can host a farmers market within their site because of GreenThumb's operating rules. Thus, the line between urban gardeners as hobbyists and urban farmers as entrepreneurs is sometimes less than clear. One activist took issue with the whole language of "urban agriculture," arguing that it reifies old divisions between the formal and informal economy:

> We don't really talk about urban agriculture. We'll say, "farms in the city." . . . For *me*, [the language of] urban agriculture is . . . trying to . . . legitimize this formal economy and say this [informal economy] isn't viable. Urban farms and community gardens are a kind of return to this civic urban agriculture, this informal economy, where the quality of life, physical activity, fresh air, local produce, connecting with people, having conversation while being outside, touching dirt. Like all of those things are valued as much as growing a tomato and saving two dollars and eating it. . . . It's all weighed in the same way. (respondent 37)

Just as small-scale rural farmers struggle to make profits due to the cost of inputs and land, so, too, do these urban agriculture groups face challenges in keeping their programs viable over time. Particularly in the face of an economic downturn, public, foundation, and donor monies are a scarce resource for which nonprofits must compete. While the income from produce sold is an important revenue stream, it is not enough to pay staff and cover the cost of operating a nonprofit. One urban agriculture business venture reported that although they were able to pay their head farmer a living wage, the founding staff members were not yet paid a regular salary for their work after several years in operation. In contrast, one group that helped to incubate a farmers market in a low-income neighborhood, which included numerous urban producers selling their goods, prided itself on its long-term viability even without a wealthy customer base. Lack of traditional sources of capital from private sources or public entities has driven innovation in alternative and crowd-sourced funding. For example, the website ioby.org (stands for "in our back yard") began in New York City as an online microphilanthropy and crowd-resourcing organization in support of urban environmental projects. As of 2011, ioby noted that 26 percent of their projects focused on gardens and another 7 percent on urban farms, and 17 percent of projects were food related (ioby 2011). Another prominent site for crowd-sourced funding is Kickstarter.org. These platforms are often used by informal, grassroots groups and individuals on particular projects and campaigns; they are not sufficient substitute revenue streams for the ongoing costs of running more formalized nonprofits.

Another important social benefit of urban agriculture is its use as a form of youth education. Urban farms and gardens are touted as sites for hands-on learning about food webs, health and nutrition, and ecology. New school gardens are being developed every season, with support from programs like Grow to Learn, which is a public-private partnership of DOE, GreenThumb, and GrowNYC created in 2010. Rooftop greenhouses offer a means for extending the growing season or creating growing space in schools without viable land at street level. The nonprofit New York Sun Works has helped schools to build and install fourteen greenhouses on school grounds since 2008 through its Greenhouse Project (NY Sun Works 2012). These school growing sites are being incorporated into classroom curricula, through the work of groups like the Greenhouse Project, Growing Chefs, and Garden to School Café. As part of the school reform and charter school movements, new schools, such as the Brooklyn Academy for Science and the Environment and the Green School, use the environment as an organizing principle throughout their institutions.

Agricultural work can also enable and benefit from extension services, adult/continuing education, peer-to-peer learning, and the application of local/traditional knowledge. Numerous interviewees cited the formative influence of Cornell Cooperative Extension officer John Ameroso, who had worked as an agronomist in New York City since 1976. Ameroso was involved in the development of almost all of the larger-scale urban farms in New York City, including ENY Farms!, Added Value's Red Hook Community Farm, the community garden at Floyd Bennett Field, and the garden at Riker's Island prison—and he helped create the City Farms program of Just Food. His legacy was highlighted in an extended profile in *The New York Times,* calling him an "urban farming pioneer" (McMillan 2010). Cornell Cooperative Extension offers services including soil testing for contaminants and educational curricula for gardens. Similarly, NYBG and BBG provide community garden extension services through the Bronx Green Up and Brooklyn Greenbridge programs, respectively. They also provide adult and youth education programs, including courses with continuing education certification in horticultural topics. The nonprofit Just Food provides numerous volunteer and compensated peer-to-peer learning programs that were often referenced by respondents as crucial to growing urban agriculture citywide. Just Food's City Farms workshops are taught by and for gardeners, with stipends provided to the instructors, and the Community Chefs program trains volunteers to provide cooking demonstrations on fresh, healthy eating at farmers markets and community gardens citywide. Farm School NYC was incubated by a collective of urban agriculturalists, including Just Food, with an aim of providing comprehensive training and certification in all aspects of urban agriculture. Farm School began organizing in 2007 and its first class began in 2011; it is rooted not only in technical knowledge about urban farming, but also a commitment to a community-based and social justice approach. Brooklyn Grange Rooftop Farm partners with a social service nonprofit focused on African refugees, offering them farm internships where they can pass on their agricultural knowledge. Outside of structured programs, gardens and farms are sites where grower knowledge—often rooted in cultural and ethnic traditions—can be shared. Overall, New York City's agricultural movement demonstrates strong professional and personal networks among individuals and organizations involved in community-based agricultural training.

This overview was not meant as an exhaustive history of community gardening and urban agriculture in New York City. Rather, it was an outline of some of the key moments in the post-1970s civic movement and the way in which GreenThumb, other city agencies, and local nonprofits helped to navigate the garden crisis of the 1990s. The memories and legacies of the 1970s founding era and the 1990s crisis thread through current land jurisdiction, policy debates, and institutional structures. Starting roughly in the mid-2000s, there was a new wave of urban agricultural practices in New York City as well as a growing local food movement. No single chapter could capture the diverse physical sites, innovative planting practices, and novel organizational forms that have developed and continue to develop as part of this evolving movement. However, I have aimed to identify the main narratives that were at play during the period 2007–2011. The constituencies involved and the storylines deployed ranged from an interest in education and employment to a focus on healthy food consumption to a commitment to food justice. Understanding the overlaps and distinctions between the prior wave of community gardeners and the newer wave of urban agriculturalists helps shed light on the visioning and planning efforts that will be described and analyzed next.

6

CITY OF FARMS

Cultivating Urban Agriculture through Food Policy Visions and Plans

Clearly, there are diverse threads within the community gardening, urban agriculture, and local food movement in New York City. Given that diversity, *a local and regional food systems framework* offered one way to bridge divisions within the movement and work toward comprehensive change to food production, processing, distribution, consumption, and postconsumption. Public officials and coalitions of activists have both used the concept of the local/regional food system. Within the initial void created by City Hall's lack of engagement, other political figures in the municipal government seized the opportunity to advance food policy agendas for New York City. Most notably, city council speaker Christine Quinn and Manhattan borough president Scott Stringer created food policy plans with wide-ranging visions as well as immediately actionable steps. Moreover, both were candidates for public office in 2014 and saw that embracing food and agriculture could help court certain constituents and identify them as "green" candidates.[1] Thus, we see that the articulation of food visions and plans are both discursive and political acts.

Subsequent to these planning efforts, civic groups and residents contested the complete absence of food, agriculture, and community gardens from PlaNYC, so much so that the 2011 update to the plan included a brief crosscutting section on food and a specific initiative on enhancing urban agriculture and community gardening citywide. Developed after the 2008 economic crisis, PlaNYC 2.0 did not include any capital commitments to urban agriculture or food systems, but it did make a number of programmatic goals and policy recommendations. It drew upon specific proposals from the prior plans of Stringer and Quinn but proposed actions that were somewhat more circumscribed in scope. Nonetheless, advocates saw the incorporation of food issues into PlaNYC as a symbolic and political achievement. How these recommendations will be carried forth into future administrations remains to be seen, but it is clear that food and, to a lesser extent, urban agriculture now have an expanded presence in the municipal policy arena.

The Local/Regional Food Systems Framework

Viewing food through a systems framework provides a large "tent" under which many diverse actors can situate themselves. Via this approach, urban agriculture is just one component of the local and regional food system. Recent civic coalitions interested in food issues in New York City have organized around a food systems approach. In part, this is because of the complexity of the challenges that the food system presents, from access to affordability to the need for awareness and education. An activist said, "Really thinking about food leads you directly to systems thinking" (respondent 31). Similarly, a researcher noted, "Focusing on regionalizing food systems . . . is a way of addressing multiple components of the problem at once" (respondent 3). One activist saw the different strands of the food movement as presenting an opportunity for collaboration:

> There was a group of people working on sustainable agriculture issues in New York City and that wasn't a very well-known movement at the time and . . . there were obviously a lot of organizations doing anti-hunger work in New York City. And so there were these two huge problems where farmers were having a really hard time connecting to markets, they were going out of business, we were losing farmland around the region, and at the same time people in New York City didn't have access to good food. So connecting the dots would be a really good way to solve everybody's problem. . . . Bringing anti-hunger and sustainable ag to the same table—I feel like that happens pretty regularly now. . . . But that really wasn't necessarily the case in the early or mid-'90s. So we've come a very long way. (respondent 60)

Yet coalition building around these diverse areas is not always easy, a policy analyst claimed: "It's a broad umbrella when you talk about food policy, which can be good. But there can be tensions of those coalitions around, I'd say, hunger advocates and nutrition folks and then people who are more into sustainable ag and a vision for upstate-downstate connections—I don't actually think hunger people care about that" (respondent 39). Despite the challenges, a number of new coalitions focused on the food system in New York City emerged and operated during the 2000s.

The Brooklyn Food Coalition (BFC), founded in the mid-2000s, was a coalition seeking to build a local food movement—with roots in Brooklyn but a reach that was citywide, region-wide, and even global. BFC hosted two large conferences in 2009 and 2012 that included hundreds of workshops and were attended by thousands of participants. At their 2012 conference, founder Nancy Romer identified four levels of action that were needed to build that movement: personal changes in how we eat; community projects that are inclusive; policy and legislative actions with elected officials; and alliance building between nonviolent social movements. Her closing plenary laid out this broad vision and closed with a call to action: "We organized this massive conference to bring the food democracy movement together. . . . We hope you will join the Brooklyn Food Coalition, yes. But most important, we hope that each of you will identify yourself as a food movement activist. To make recognition of what a better food system would look like. We can see the road we need to get there. Let's get on that road and let's do it together!" (Romer 2012). The BFC website, email lists, and meetings provided further fora for social networking and organizing, both through neighborhood-based work and through cross-cutting themes of food policy, school food, workers' rights, research, and more. A policymaker noted that the number of attendees at the 2009 conference was a sign of the health of the food movement in New York City and did not go unnoticed by public officials. However, as of the summer of 2016, the BFC website listed the coalition as "on hold."

Similarly, the Food Systems Network NYC (FSNNYC) was a coalition of individuals committed to improving the regional food system with a focus on health, nutrition, economy, upstate-downstate connections, and "the support of a vibrant food and farming economy" (respondent 31). The coalition's loose collaborative structure was a deliberate choice by the founders, who realized that their members could not always represent their organizational affiliations in their advocacy work but could participate as individuals. FSNNYC's approach was to "make the food movement more ecumenical"—meaning more collaborative, networked, and broad-based (respondent 38). In addition to various subcommittees, the coalition had monthly open networking meetings, where experts from different areas made presentations and members and the public attended to learn and meet one another. The members engaged in policy advocacy through petitions, meetings, and memo writing. With funding from the Kellogg Foundation and in partnership with the DOHMH, the coalition developed a program focused on neighborhood food planning and community-based mapping, entitled FoodAction. Thus, like the BFC, FSNNYC facilitated individual education, supported neighborhood-level action and programs, and engaged in citywide, regional, and federal policy discussions. In terms of tone, issues of food justice were less prominent in the language of the FSNNYC than of the BFC, though the mission and core values of the group certainly mentioned these issues.

It is important to note, however, that as of 2016, neither of these local coalitions was operating. These loose coalitions can rise and fall with particular leaders and funding opportunities—but their impact on broader movements and individual minds can have ripple effects beyond the life of their operation. At the same time, other coalitions and networks related to community gardening continue to persist, such as NYCCGC and Farm School NYC. National organizations and networks such as Food Tank and the Fair Food Network continue to share best practices and advance policy dialogues about food systems change across the country. Meanwhile, within the New York City municipal government, previous silos of topical focus have also begun to be broken down in support of improving the food system. This has occurred, in part, through the creation of the food policy coordinator (FPC) position and the Office of Long-Term Planning and Sustainability (OLTPS), which was responsible for coordinating PlaNYC. I turn next to examine municipal vision and policy documents released between 2007 and 2011 (*Food in the Public Interest, FoodNYC, FoodWorks,* and PlaNYC 2.0's chapter on food). All of these deployed a systems framework, wherein goals, recommendations, and strategies were organized by the stages in the food system: production, processing, distribution, consumption, and postconsumption.

Food in the Public Interest and *FoodNYC*: Visionary Plans as Political Strategy

Completely outside of the PlaNYC processes, Stringer demonstrated an interest in issues related to the local food system. Starting with his "Go Green East Harlem Campaign" in May 2007, Stringer began to provide small amounts of funding and support for farmers markets, community gardens, and community chefs in East Harlem. The project also led to the creation of a "Go Green East Harlem" cookbook (Stringer 2009, 2012). In speaking at the Brooklyn Food Conference, Stringer said that his intention with this initiative came from a social justice perspective:

> For me, it started when I got elected Manhattan borough president a number of years ago, when I realized . . . "I'm going to represent some of the *wealthiest* people in this city, in Manhattan." Think about the great concentration of wealth and people doing very well. I wanted to focus on communities that actually weren't doing so well. I wanted to start an initiative on sustainability in a community—neighborhood that really needed some help and that was *el Barrio* in East Harlem. Part

> of what I wanted to do there was build a grassroots, bottom-up movement. . . . With community-based planning and some strategy we could create an atmosphere where we could plan for more parks space, plan for more open space, create a sense that the community did not have to be unhealthy, did not have to have food deserts, did not have to have a neighborhood that was struggling with asthma, diabetes, and other unhealthy choices. (Stringer 2012)

Stringer began his engagement at the neighborhood level, focusing on areas of need within his jurisdiction before expanding his scope citywide and beyond.

As he began to scale up his engagement, Stringer organized two conferences with university, nonprofit, and community-based allies. Many activists acknowledged these conferences as being crucial moments of coalescing for the citywide food movement. The first conference, entitled "The Politics of Food," was held at Columbia University in November 2008. Activists described the planning and implementation of this conference as collaborative and inclusive. The event itself was viewed as an opportunity to network and share ideas among the various threads of activism in the food system. A policymaker described the motivation for the event from the perspective of the borough president's office:

> No one was really talking about [food] in a broad sense. And I think the borough president wanted to start that dialogue, which is something that I think borough president's offices are poised to do. They don't necessarily do them. But Scott Stringer as an individual is someone who is invested in process—and an inclusive process that brings different voices to the table and [he's] more of a bottom up kind of guy rather than a top down. . . . And so the idea was to really create a space, while acknowledging that we weren't experts. . . . He started to see that . . . there needed to be a larger dialogue about how food fit into these larger poverty and justice issues that we were trying to approach. (respondent 39)

Approximately six hundred people, including decision-makers, activists, community residents, and scholars, participated in the conference (Stringer 2009).

The report *Food in the Public Interest* was released in February 2009 as a result of the ideas that were generated in the conference and then further refined by Stringer, his staff, and a steering committee; the report thanked seventy-one individuals and more than forty institutions in the acknowledgments. The document framed food and agriculture issues around health, environment, and economic concerns. It made recommendations in several domains: hunger; urban and regional agriculture; food distribution; economic development; food and nutrition education; and steps toward implementation. Homing in on the urban and regional agriculture section, it framed the issue around New York State as an agricultural producer and the need to enhance upstate and downstate connections in order to strengthen the regional foodshed. At the same time, it acknowledged the cultural benefits of local food production and called for greater assessment of the challenges facing, as well as opportunities for, urban agriculture. The specific recommendations called for research on the regional foodshed and land available for urban agriculture; changes to New York City agencies' food procurement policies; policies that target low-income food deserts; efforts to promote local food production; and the development of educational information and curricula about urban and regional agriculture (Stringer 2009, 8). The emphasis on research and information gathering is not surprising, given that some of the individuals who cosponsored the conference were affiliated with the Columbia University Urban Design Lab.

With respect to implementation, *Food in the Public Interest* noted the complex, multiscaled, and networked nature of food systems problems—making them particularly difficult to address by means of narrow policy solutions, stating, "There is clearly no silver bullet to address this multi-faceted issue" (Stringer 2009, 12). The report acknowledged that historically, the federal government and the private sector have played a leading role in food policy, but called for increased engagement from municipal governments, nonprofits, and the public. Noting the role that Bloomberg and Quinn played in policies supporting healthy eating, it called for more action through the office of the FPC as well as the creation of a Food Policy Council, comprised of both public officials and civic representatives, to develop a comprehensive food systems plan. It also called for policy advancement via a food charter and a food assessment, including indicators of success and a food and health report card. Beyond that, it used the language of social movements, calling on us all to "develop a critical mass, a movement to effect change" (Stringer 2009, 12). And it made a normative call for attention to the underserved, marginalized, and low-income residents.

A second conference focusing on the links between the food system and climate change ("Food and Climate Summit") was held at New York University in December 2009. This event was organized in concert with the United Nations' Copenhagen climate talks, involved video feeds and live tweeting from the global meeting, and engaged quite a few rural/upstate farmers in the discussion. This, too, was followed by a Stringer-led report, entitled *FoodNYC*, which was released in 2010. The primary focus of the document was on health issues, with other dimensions of climate change mitigation and adaptation and sustainability layered on top. The specific recommendations addressed urban agriculture; regional food production (this time decoupled from urban); food processing and distribution; new markets; the procurement of regionally grown food; education; food waste; plastic bottles; food economy; and a proposed new department of food and markets (Stringer 2010).

This report was situated in the context of other "current policy foundations" both in New York and elsewhere (Stringer 2010, 7). First, *FoodNYC* made reference to the policies, programs, and initiatives of other cities. This served to help legitimize the claims and recommendations made within the report because they were pursued in other locales and positioned the work within a frame of New York City's global competitiveness. Second, beginning with its name, the document was clearly in close dialogue with the mayor's PlaNYC. The absence of food systems from PlaNYC and the city's greenhouse gas inventory was explicitly noted: "Food has not been included in these reports. As of yet, the City has not stated its commitment to creating a sustainable food system nor has it done the baseline research needed to determine the most important sources of emissions and other environmental impacts of New York City's food system" (Stringer 2010, 12). The document was written with an aim toward influencing policy conversations in a number of arenas, including the 2011 update to PlaNYC. *FoodNYC* explicitly recommended that food be added to PlaNYC and called for the creation of a municipal Department of Food and Markets, referencing the precedent of the comprehensive food policy of San Francisco (Stringer 2010, 34). Reflecting on the impact of *FoodNYC* on PlaNYC 2.0 and *FoodWorks,* an interviewee said:

> I think it's made a difference. I think we were the first people to really start the dialogue. I think we did change the paradigm. I don't think without our conferences there would have been a food chapter in PlaNYC.... I think that we've been successful in shifting this idea about the role of food systems and what that means in the city. And I don't think *FoodWorks* would have happened if we hadn't done *FoodNYC.* And as a result a lot of legislation that came out of [FoodWorks] was exactly what we had recommended: the inventory of the land, the stuff around [local agency food] procurement. (respondent 39)

In terms of its focus on urban agriculture, *FoodNYC* adopted the PlaNYC timeline, targeting the year 2030 with the goal "Establish food production as a priority in New York City for personal, community, or commercial use by the year 2030" (Stringer 2010, 2). Specific recommendations were "Assess Land Availability and Suitability for Urban Agriculture"; "Create a Citywide Urban Agriculture Program"; "Ensure the Permanence of Community Gardens"; and "Facilitate the Development of Rooftop Agricultural Greenhouses" (Stringer 2010, 9–10). The first two proposals were carried over from the prior *Food in the Public Interest* report, with additional mimicking of PlaNYC models. Indeed, the urban agriculture program referenced MillionTreesNYC as a precedent:

> The Mayor should establish a citywide Urban Agriculture Program to support the creation of food growing spaces. The program should be similar to Capital Growth, London's campaign to plant 2,012 growing spaces by the 2012 Olympics by connecting people to land, providing funding, and offering practical guidance on how to grow food. New York City's Urban Agriculture Program should be modeled on and keep pace with "Million Trees NYC," an initiative announced by Mayor Bloomberg on Earth Day in 2007 to plant one million trees throughout the city by 2030. It would thus be operated through a public-private partnership and draw a large volunteer base (Stringer 2010, 9).

The addition of community garden protection to Stringer's agenda reflected the growing dialogue around that topic among civic actors and public officials after the 2010 expiration of the attorney general's settlement (discussed in chapter 5). The inclusion of a recommendation related to rooftop greenhouses reflected discussions around policy barriers and challenges facing rooftop farmers. Finally, this section continued the practice of positioning New York City in the context of the work of other cities, mentioning Vancouver, Portland, London, Brisbane, and San Francisco.

In order to appreciate Stringer's political role and the impact of these documents, it is important to understand that the position of borough president is something of a remnant of New York's preconsolidation history as several distinct cities.[2] As such, the position has little formal authority but remains a visible figure in New York City politics (Eichtenthal 1990). Several respondents, even ones who worked with Stringer, commented on the limitations of this public office. A researcher said, "To most people unfamiliar with civics, [*FoodNYC*] was the food plan for New York City. My colleagues from outside of New York City said, 'Wow, you guys have a food plan and it looks really great.' Of course I had to explain that it's not an official food plan. Sorry to be so cynical, but something issued by the borough president, it's kind of a vestigial organ of city government. It only has meaning to the extent that the borough president is willing to make it have meaning" (respondent 12). While officials within the mayoral administration critiqued the Stringer plans for not being "actionable," his staff believed that describing a progressive vision was part of the role of the office. He could use the power of the "bully pulpit" to be a "voice for the underdog" and help to push the whole policy dialogue leftward (respondent 39). Allowing these plans to be visionary documents, particularly given the circumscribed authority of the office, was part of the political strategy. Yet a high-ranking public official within a city agency disputed the effectiveness of this approach. She viewed Stringer's strategy as more about "making a name for himself" than about seeking to affect real change. She noted that in order to implement many of the recommendations, Stringer would have to let go of some of the personal credit and work through the slow process of building approval and support with the mayor, the council, and city agencies in order to institutionalize change (respondent 50).

Stringer continued to claim food policy as part of his agenda as he positioned himself for an electoral run—first running for mayor, then deciding to run for city comptroller after November 2012. In 2011 he released the report "Green Vegetables, Red Tape," critiquing the regulations surrounding farmers markets and advocating for more community-based farmers markets. Although the focus of the document was on farmers markets, it also had a series of recommendations about fostering urban agriculture by conducting an assessment of the potential for urban agriculture, creating a citywide urban agriculture department, and working to preserve community gardens. Thus, it maintained the same recommendations related to urban agriculture that were proposed in *FoodNYC* and continued to consolidate his food platform (Stringer 2011). In 2012 Stringer spoke about his track record of organizing and engagement on food issues at the Brooklyn Food Conference. In his remarks, he discussed the "economic advantage of the urban food agenda," sounding more like a supporter of the neoliberal competitive city, while elsewhere in those same remarks he sounded like a progressive voice of the left championing a food movement (Stringer 2012). This, perhaps, reflected the status of a man at a political juncture, someone who saw himself as a voice of the underdog but with an eye toward the highest public offices in the city.

FoodWorks: Food System Dialogue Enters the Legislative Arena

Former city council speaker Christine Quinn also developed an interest in local food systems as part of her policy agenda. While Quinn had shown long-time personal engagement with her local CSA and with community gardens and gardeners in her district, she had not, prior to 2010, developed a full-fledged formal food policy. Her prior policy engagements had focused on combatting hunger and promoting job development via the FRESH program and supporting farmers markets, food stamps, and community incubator kitchens; and she became involved in the community garden rule revision after the expiration of the settlement. In particular, her work was rooted in a public health perspective: "Food was something that the speaker had already been working on for several years, since she had come into office. And she had been coming at it from more of a hunger and health perspective, which is where a lot of people started out in the food policy world . . . framing it in terms of obesity and hunger and diabetes. And so she had done a lot of initiatives on food stamps and getting a Food Policy Coordinator office" (respondent 8). It was Quinn's young staffers—influenced by the growing national discussion around local food described in chapter 5—who first helped to fully articulate and develop a broader approach to food policy beyond obesity and hunger that included a "systems perspective and looking at . . . where there are points of insecurity or little fissures in this system that are creating—not just the health outcomes that we're seeing—but also economic and environmental [outcomes], because you can't really disentangle those things" (respondent 8). Through a food systems framework, Quinn's office was able to address issues across the entire spectrum of production, processing, distribution, consumption, and postconsumption (NYC Council 2010).

FoodWorks was launched publicly as an initiative of the speaker in December 2009, and the plan was issued a year later in November 2010. According to some respondents, the process of its development was not completely transparent or inclusive. Unlike the Stringer documents, which were shaped by broadly attended conferences, *FoodWorks* had a hand-picked set of nine advisors who gave input to the plan (Stringer 2009, 2010; NYC Council 2010). Indeed, by one of these advisor's accounts, it was he who brought the idea of a food systems plan or vision to the speaker and her staff. Quinn's staff conducted dozens of one-on-one interviews with experts and advocates, but these individuals were not brought together in a deliberative setting. Some activists critiqued the planning process as exclusive and questioned its representativeness of diverse New York City constituencies:

> [I was] very, very upset with it, because when [Quinn] announced the unveiling of her plan and she brought up to the stage, the people who were behind the plan in terms of helping her, there was not one person of color. . . . And so . . . I brought it up . . . . "People, let's get real. How are you talking about food when the most impact that food has is [on] low income neighborhoods, neighborhoods of color? And so if we're going to talk about solutions, we've got to be at the table." . . . So I was very disappointed in the fact that all these people, all these non-profit organizations, and state and city agencies were called. And it's like, "Okay, where are we in this?" (respondent 29)

When probed further, the interviewee identified "unintentional racism" as the reason behind white leaders surrounding themselves with white "experts" with whom they are comfortable and share a common background (respondent 29). Because of the deep commitment to procedural justice on the part of many urban environmental and food justice groups, one activist claimed, "If it's not inclusive, it's flawed" (respondent 33).

It is instructive to examine *FoodWorks* in the context of mayoral politics. First, a source familiar with Quinn noted the impression that the 2009 Brooklyn Food Conference made on Quinn. Speaking at the conference and expecting to see a few dozen to a few hundred activists, Quinn was shocked to see nearly one thousand participants inside on a Saturday talking about all aspects of the food system. This was a catalyzing moment for the speaker when she realized that this could be an important constituency for her and that this movement had "legs." Second, an interviewee familiar with the Stringer conferences and plans felt that *FoodWorks* built on this prior work, without explicitly acknowledging it. As both were at one time mayoral candidates, there was pressure to delineate and claim leadership in certain policy arenas. Finally, interviewees knowledgeable of the internal decision-making process in Quinn's office said that *FoodWorks* was fast-tracked in order to precede the release of PlaNYC 2.0 in April 2011. This was done knowing that Bloomberg's signature initiative would include some mention of food; and there was a desire to claim the intellectual turf around food policy for Quinn and *FoodWorks*. This maneuver had to be handled delicately, however, as Quinn and Bloomberg had to maintain a positive working relationship as speaker and mayor; and candidate Quinn sought a strong Bloomberg endorsement.

FoodWorks was an eighty-six-page document, ambitious in its aims and networked in its approach. It situated the issue of food policy and food systems in a centuries-long timeline about food production and consumption in the introduction. It included recommendations that were beyond the authorities of the city council and it touched on geographies and policies beyond the municipal scale—including state agricultural regulations, federal food stamps, and farm subsidies. And it involved multiple sectors, reflecting the speaker and her staff's interest in networked governance. A policymaker said, "Public policy has been moving toward this networked approach. That it's not just about government doing things and having a role, but it's about the way that government and private sector, and the not-for-profit sector can work together to achieve common goals" (respondent 8). In its introduction, the report framed the issues around the economy ("seizing economic opportunity"), the environment ("improving environmental sustainability"), and health ("improving public health"), rather than an explicit focus on social justice. *FoodWorks* also used the framing of population growth that was the departure point for PlaNYC but nuanced it, given the complicated relationship between food systems and population growth, as follows:

> Food systems have changed throughout history to support the evolution and economic growth of societies. Today we are once again confronted

> with the need for additional change to the food system. Our national food system evolved to support a rapidly growing population, and it has allowed us to feed more people than ever before. Yet, that evolution had unintended consequences. . . . As New York City is expected to add nearly one million new residents in the next two decades, we must identify ways to move from an unsustainable food system to one that promotes health, environmental sustainability, and a thriving economy. (NYC Council 2010, 2)

Finally, a certain amount of interurban competitiveness pervades the text, with references to making New York City a "leader" in food policy and food systems change (NYC Council 2010).

Related to food production, *FoodWorks* set two primary goals that span the rural-to-urban gradient: "preserve and increase regional food production" and "increase urban food production" (NYC Council 2010, 18).[3] The proposals within these goals echoed many of the recommendations from earlier Stringer documents and conferences, such as building a permanent wholesale farmers market, expanding and supporting farmers markets and CSAs, and expanding the acceptance of EBT at farmers markets. Focusing on urban agriculture, the strategies aimed to "better use existing space for urban food production" and to "restore food and horticultural knowledge." The emphasis on space continued the discussion around scarcity of land in the developed city and around proposals related to protecting community gardens. It also began to address the "lack of data" critique with proposals focusing on collecting data on urban farms citywide and creating a searchable database of city-owned land. A suite of measures related specifically to rooftop agriculture: identifying city-owned buildings that could potentially host rooftop farms, making changes to building requirements and waiving floor area ratio requirements for rooftop greenhouses, changing green roof tax credits to encourage food-producing roofs, and changing water rates for rooftop agriculture sites (NYC Council 2010). Indeed, the number of proposals focusing on rooftop agriculture led one interviewee to critique the overemphasis that these sites received in the document.

In order to show progress on this ambitious agenda, Quinn's staff set about developing legislation, resolutions, and recommendations for policy changes that could be effected immediately at the municipal scale. The city council passed local laws in August 2011, targeting different points in the food system. To begin to address the question of how much land is potentially available for urban agriculture, local law 48 required the Department of City Administrative Services (DCAS) to maintain an online database of all publicly owned property in the city, with information about its suitability for agricultural uses.[4] Local law 49 excluded rooftop greenhouses from contributing to height restrictions under zoning and building codes. As a first step in working to change public agencies' food procurement practices, local law 50 required DOE and other agencies that procure food to report on the amount of regionally sourced, New York State foods. Local law 51 focused on reducing packaging waste by developing guidelines for packaged food purchased by the city. Finally, local law 52 required OLTPS to gather and report on information about New York City's food system and to develop an annual food system metrics report.[5] All of these pertain to food systems, but laws 48, 49, and 52 have the most potential impact on urban agriculture. While many respondents commended the progress, even those internal to the *FoodWorks* process noted that these local laws are only a first step. Much more work remains to be done to close loopholes, strengthen enforceability, and build on this early momentum.

Despite these first legislative steps and the excitement brought about by this public attention to food policy, some activists critiqued *FoodWorks* for not doing enough to support actual on-the-ground programs and saw it as a missed opportunity to do more. For example, one nonprofit manager noted that her program operates at the unique intersection of urban agriculture, job training, and the strengthening of healthy communities that is called for in the plan, but she saw no increase in funds received from Quinn's office or from elsewhere in the council. A public official considered the local laws and resolutions as something of an "unfunded mandate" (respondent 6). Without commensurate budget increases to the agencies involved in supporting urban agriculture like DPR and DSNY, how would all of the proposed innovations across the food cycle be carried out? These challenges bring us back full circle to the issue of the balance of power between the mayor and city council the limits on the authority of the council, and the need for mayoral engagement in order for initiatives to reach full fruition. Hence, activists were excited about the potential for Quinn's leadership on these issues if she had been elected mayor, viewing her activities while speaker as a sign of more things to come.

Food, Agriculture, and Gardens Make Small Inroads in PlaNYC 2.0

There was a clear public response to PlaNYC's failure to mention gardening, farming, agriculture, or food in any form: focused advocacy and community organizing to inform the 2011 PlaNYC update.[6] Individual residents, civic groups, and coalitions critiqued PlaNYC substantively for what it lacked in content and procedurally for the lack of public participation in its development. An activist noted, "If you look at the plan, it did not talk about food. And so many of the listening tours brought that up. You're talking about housing, the trees, jobs. But you didn't talk about food. And that's critical. How are you going to feed all these people in the anticipation of millions of people? How are you going to address that?" (respondent 29). Reflecting on this absence, some interviewees noted that there is no single, municipal agency that covers food issues for New York City. While that may be the case, there are several key agencies that offer long-standing support to urban agriculture and community gardens in the city. In particular, GreenThumb (under DPR) and NYCHA are heavily involved; and HPD, DCAS, DSNY, DOE, and DOT all have roles in supporting this system, as described in chapter 5. Moreover, the issue of interagency coordination should not prevent policy innovation, as PlaNYC and OLTPS were created to seed and ensure just such collaboration. One activist felt that the absence was due to the prior focus of PlaNYC on climate change and on the environmental dimensions of sustainability at the expense of other dimensions that might be highlighted through progressive food policy, such as the creation of green jobs. Scholarly literature has noted this leaning of PlaNYC toward the environmental, at the expense of equity or social justice, as well (Finn and McCormick 2011; Rosan 2011). Similarly, another advocate felt that the omission was directly related to personal perceptions of social justice, because city politicians who do not experience hunger or poverty may fail to see the food systems as a critical issue.

Once local law 17 ensconced sustainability planning in the charter and required an update to the plan every four years, advocates began targeting the update to the plan to include food, farms, and gardens in its scope. For example, the FSNNYC organized a policy memo response to PlaNYC, called "Food for the Future," that was sent directly to the mayor and other key decision-makers. The authors aimed for this memo to be as quantitative as possible, in order to appeal to the authors of PlaNYC: "I kind of understand the way the city thinks . . . the perspective of PlaNYC. They want everything to be measured: controllable, measurable, defined outcomes, progress against plan. All those good, rational bureaucratic things. . . . What we tried to do was to make [the Food for the Future memo] as quantifiable as we possibly could. . . . Because quantification is going to be the kind of thing that the authors of the plan get" (respondent 38). Concurrently, community gardeners contacted GreenThumb and other city agencies to voice their discontent about the absence of gardens from PlaNYC. And they showed up en masse to outreach meetings organized around the development of PlaNYC 2.0.

OLTPS sought to give the public and civil society groups broader roles in the planning process surrounding PlaNYC 2.0. This served the dual function of both responding to prior criticism on procedural grounds as well as potentially leveraging the contributions and resources of outside entities in dire financial times. A public engagement process was developed to help shift PlaNYC "from an elite plan to a democratic plan" (respondent 49). Borough-based and thematically focused outreach sessions were held at public venues and City Hall. And officials recalled being surprised by the number of attendees at the sessions on food and agriculture. Asking the public for input—while framed as a way of gathering ideas—was also a way of developing buy-in for strategies that planners had already developed. In addition, these face-to-face meetings were supplemented with an online tool that the deputy mayor's office developed. Entitled "Change by Us," this website was designed to gather ideas, route people to municipal resources, and connect groups to each other; it was envisioned as a way of virtually helping to spur civic engagement in neighborhood and environmental issues.

Despite the public engagement, PlaNYC 2.0's food goals were more circumscribed in scope than those in *FoodNYC* or *FoodWorks*. First, the issue of food was framed, limited, and bracketed by caveats in the final text of the plan:

> Healthy, sustainable food systems are critical to the well-being of our communities and central to our ability to accommodate a growing population. Yet food presents a unique planning challenge; unlike sewers or streets, much of New York City's food systems infrastructure is privately owned and shaped by the tastes and decisions of millions of individual consumers. These complicated and inter-related subsystems aren't easily understood or influenced, even by concerted municipal interventions. Furthermore, many of food's most significant climate and environmental impacts are associated with food production, most of which takes place outside the city, and shaped by federal policy. (City of New York 2011a, 164)

Second, during the plan's development, the mayor's office sought to identify goals that they knew the city had the jurisdiction, authority, and resources to achieve. This limited ambition resulted from the fact that PlaNYC 2.0 was released in a completely different fiscal climate than the first version of the plan. There was no budget surplus with which to work, and many municipal agencies had experienced across-the-board budget cuts of 30 percent. City Hall and agency staff were tasked with coming up with goals that could be met with budgets that were already in hand. One decision-maker commented on PlaNYC 2.0 as focusing on incremental, no-cost changes: "You don't mention huge capital dollars. You need to *tweak* things. You need to nip and tuck" (respondent 52). Participants involved in the planning process were asked to identify targets related to urban agriculture and community gardening *that they knew they would meet regardless of the plan.* According to one public employee:

> It basically was right before they put the draft out. . . . I went in this meeting with [City Hall officials] and they were like, "We need to put in something about food production and community gardens in PlaNYC. What can we put in? But please make sure that you understand that there's not going to be any additional budget for it at all and it has to be something that you can definitely do with your existing funding and nothing else." I was like, "this is the most cynical exercise I could possibly imagine." But I was still happy that they were going to be in there. So I said, "How about we work with Just Food? Start a few more farmers markets. Let's make it a reasonable number like what we would probably do anyway. Like five. We increase the population of gardeners by a certain percentage cause that's what we have to report to OMB anyway. So it's what we're already trying to do. And we create more access to community gardens in areas that are underserved by community gardens, like Queens." Literally, that went into PlaNYC—almost word for word. The only thing they added was the . . . searchable database of property available for community gardening. That didn't come from me. (respondent 6)

Because resources were not committed through PlaNYC 2.0, some advocates felt that the city was not ready to engage seriously in food policy.

Others acknowledged that just beginning to write about food, agriculture, and gardens was a step in right direction, albeit a small one. From no mention of these topics in the first iteration of the plan, PlaNYC 2.0 mentioned community gardens/gardening more than thirty times. These references occurred not only in the short, two-page cross-cutting theme on food that was added at the end, but also in the core chapters on parks, public space, and brownfields. Other cross-cutting themes added to the document included public health, natural systems, green building, waterfront, economic opportunity, and public engagement—and an entire core chapter was added on solid waste issues (City of New York 2011a). One respondent noted:

> You know, the fact that . . . somewhere you can go to a website and show that New York City supports a food policy for the city, supports having an open space within walking distance of every resident in the city, that supports planting trees and supports any efforts to prevent erosion and pollution of the waterways. That's got to be a good thing because that wasn't there before. We have a supportive mayor who wants . . . to be known as an environmental mayor. We don't know that the next one is gonna' be quite the same. I think it's gonna' be really hard for that next mayor to sort of put the brakes on all of this stuff and reverse it. But they could be much less supportive. They could be much more development [focused]. (respondent 58)

The public space chapter took an environmental justice–oriented approach to the goal of "Target[ing] high impact projects in neighborhoods underserved by parks." This goal included an initiative to "facilitate urban agriculture and community gardening" (City of New York 2011a, 35, 37). Following on the precedent set in the 2007 plan, PlaNYC 2.0 set numeric goals. It committed to planting 129 new NYCHA community gardens and creating one urban farm at an NYCHA site; registering 25 new school gardens per year to the Grow to Learn program and retaining at least 75 percent of registered gardens each year; increasing the number of registered GreenThumb garden volunteers by 25 percent; and establishing five new farmers markets at community garden sites, in a partnership between GreenThumb and Just Food (City of New York 2011a). While these numeric goals were easily tracked, they were also easily met, "low-hanging fruit" goals that did not involve any commitment of funds or expansion of municipal programs.

Just as *FoodWorks* picked up on frameworks, concepts, goals, and policies identified in the earlier Stringer documents, so, too, did PlaNYC 2.0 build upon all three prior planning efforts. One of the areas of overlap was the call for a public, searchable database of city-owned vacant land that potentially could be used for urban agriculture, as codified in local law 48. PlaNYC 2.0 called for the city to "survey municipal lands to identify underutilized properties that may be suitable for urban agriculture or community gardens" (City of New York 2011a, 164). This goal catalyzed an Urban Agriculture Taskforce, which includes representatives from GreenThumb, DCAS, HPD, OLTPS, the FPC, and City Hall. PlaNYC 2.0 connected to the ongoing discussions around rooftop farms, stating, "We will also review existing regulations and laws to identify and remove unnecessary barriers to the creation of community gardens and urban farms. For example, only green roofs that use drought-resistant plants are currently eligible for the New York State green roof tax abatement. Broadening this legislation to include agricultural plants could encourage an increase in green roofs and urban food production" (City of New York 2011a, 37). Other goals suggested urban agriculture as a potential use for remediated brownfield sites, through the use of protective liners and other technologies to ensure that toxins do not escape. Some connections to upstate agriculture were made through the existing Watershed Protection Program, which works with farmers to "adopt sustainable agriculture practices" (City of New York 2011a, 164). Finally, the newly added solid waste chapter included an initiative to "create additional opportunities to recover organic material" and a specific effort to "reinstate leaf and yard waste collection for composting in the city" (City of New York 2011a, 140). Reinstating and expanding municipal composting programs was one of the key policy changes called for by garden advocates and articulated in *FoodWorks*. See table 6.1 for a comparison of goals related to urban and regional agricultural production in *FoodNYC*, *FoodWorks*, and PlaNYC 2.0.[7]

TABLE 6.1 Goals, initiatives, and proposals related to urban and regional agricultural production[1] in *FoodNYC, FoodWorks,* and PlaNYC 2.0

FOODNYC

GOAL	STRATEGY / INITIATIVE	PROPOSAL / TARGET ACTION
Establish food production as a priority in NYC for personal, community, or commercial use by the year 2030	Urban Agriculture	• Assess land availability and suitability for urban agriculture • Create a citywide urban agriculture program • Ensure the permanence of community gardens • Facilitate the development of rooftop agricultural greenhouses (pp. 9–10)
Promote and support regional agriculture by connecting upstate and Long Island farms with downstate consumers, and by mapping the food grown and sourced from the region within approx. 200 miles of NYC	Regional Food Production	• Determine the capacity of the regional foodshed • Develop a state strategy for farmland and food production • Accelerate the protection of New York's farmland (p. 12)
Launch twin composting initiatives: (a) support for large-scale composting through creation of a municipal facility; and (b) support for small-scale composting through education, decentralized composting bins, and more pickup locations	Food Waste	• Eliminate barriers to food composting in community gardens (p. 26)
Educate New York City's children to become a new generation of healthy and environmentally aware eaters	Education	• Expose city students to farms and gardens—"The State Legislature should also mandate that every school has access to agriculture, be it a community garden, urban farm, or relationship with a rural farm" (p. 24)

(*Continued*)

TABLE 6.1 Continued

FOODWORKS		
GOAL	**STRATEGY / INITIATIVE**	**PROPOSAL / TARGET ACTION**
Preserve and increase regional food production	Strengthen regional food supply channels	• Reorient federal farm subsidies to support healthy, sustainable food production • Improve the New York State Farmland Protection Fund • Encourage new farmers • Build a permanent wholesale farmers market • Expand and support farmers markets • Expand the electronic benefits transfer (EBT) program and acceptance of the Special Supplemental Nutrition Program for Women, Infants, and Children (WIC) benefits at farmers markets • Expand and support community supported agriculture (CSA) (pp. 18–22)
Preserve and increase regional food production	Leverage the city's economic power to support regional producers	• Track and encourage regional food procurement • Support farmers in the upstate watersheds (pp. 22–24)
Increase urban food production	Better use existing space for urban food production	• Protect community gardens • Ensure urban farms are counted in the Census of Agriculture • Create a searchable database of city-owned property • Identify city-owned properties with roofs suitable for urban agriculture • Waive the floor to area ratio (FAR) requirements and height restrictions for certain rooftop greenhouses • Change the state green roofs tax credit to encourage food-producing green roofs • Change water rates to encourage green roofs • Streamline the green roof permit application process (pp. 26–29)

Increase urban food production	Restore food and horticultural knowledge	• Ensure garden education is available citywide • Support the development of urban agriculture technology (pp. 29–30)
Increase resource recapture in the food system	Increase residential, commercial, and governmental composting	• Establish a voluntary household composting program: "By recapturing these source-separated organic materials, we are also diverting this byproduct from consumers into a stream to use the materials as a resource for growing food" (p. 71)

PLANYC 2.0

CHAPTER	GOAL	STRATEGY / INITIATIVE	PROPOSAL / TARGET ACTION
Parks and Public Space	Target high impact projects in neighborhoods underserved by parks	Facilitate urban agriculture and community gardening	• Increase the number of registered GreenThumb garden volunteers by 25 percent • Establish five new farmers markets at community garden sites in partnership with Just Food • Register twenty-five new school gardens per year to the Grow-to-Learn program, and retain at least 75 percent of registered gardens each year • Take a full inventory of municipal land and identify properties that could be suitable for urban agriculture • Review existing regulations and laws to identify and remove unnecessary barriers to the creation of community gardens and urban farms • Plant 129 new NYCHA community gardens, and create one urban farm at an NYCHA site (pp. 37, 164)

(*Continued*)

TABLE 6.1 Continued

PLANYC 2.0			
CHAPTER	**GOAL**	**STRATEGY / INITIATIVE**	**PROPOSAL / TARGET ACTION**
Solid Waste	Increase the recovery of resources from the waste stream	Create additional opportunities to recover organic material	• To capture the roughly 4 percent of residential waste made up of leaf and yard trimmings, we will reinstate leaf and yard waste collection for composting in the city. This will create a high-quality soil product for use by city agencies and nonprofits in parks and natural resource programs (p. 140)
Brownfields	Expand the use of green remediation	Promote green space on remediated brownfield properties	• Design protective measures such as liners for state-of-the-art community gardens on remediated brownfield properties • Work with GreenThumb and the NYRP to pilot a community garden on a remediated brownfield site (pp. 57, 164)
Water Supply	Ensure the quality of our drinking water	Continue the Watershed Protection Program	• Continue our partnership with the Watershed Agricultural Council to promote sustainable farming techniques that limit the amount of fertilizer and other waste products that run into our reservoirs (pp. 81, 164)

[1]Recognizing the complexity and interconnectedness of the food system, this table focuses only on initiatives for strengthening *food production*. Thus, it does not cover issues such as distribution (including farmers markets) and post-consumption (including composting) unless explicit reference is made linking these policies to production.

Besides the specific numeric targets, PlaNYC 2.0 contained essentially no new goals related to urban agriculture or food systems and was more reserved in scope and ambition than the Stringer and Quinn efforts. However, the PlaNYC imprimatur helped to bring food issues further into the institutional fold of the administration. Building upon the local legislation to create a database of city properties through an interagency Urban Agriculture Taskforce is one clear example of how City Hall and OLTPS facilitated collaboration across executive agencies. Indeed, at the first meeting of the taskforce, many of the same activists that had been protesting the bulldozing of community gardens on the steps of City Hall in the 1990s marveled at how far the movement had come, with activists and policymakers now gathered around a table inside a City Hall conference room. Moreover, even if PlaNYC 2.0 did not bring with it extensive food policy changes, it was seen as signifying potential future directions in which City Hall and agencies might head. Many of the long-term goals were contingent: they can change with future mayoral administrations. As an executive-led initiative, PlaNYC was identified fundamentally with Bloomberg; de Blasio and other future mayors would have to establish their own, unique programs and policies:

> I don't think PlaNYC will last because it's so associated with this mayor that I think the next mayor would be hard pressed to be able to adopt it, be able to contribute new ideas to it and then still have it be associated with—whatever new ideas get put into PlaNYC would be associated still with Mayor Bloomberg, at least for the next term. So I think what's going to happen is the office [OLTPS—now MOS] will stay. I think the personalities in that office will change. And I think that the emphasis on certain issues may change. And maybe the name of the plan will change. . . . These mayors, they want to leave their own mark, which I don't blame them. And it's good for the city 'cause then these new ideas come up. But PlaNYC is a brand. (respondent 26)

At the same time, despite the need to "re-brand" PlaNYC, this same respondent felt that the way of thinking about sustainability was institutionalized enough that it could not be easily reversed: "But this is now part of government's responsibility . . . promoting a sustainable city is now part of the mission of a good administration. And so in that sense I think it will live on probably forever" (respondent 26).

This case reveals the complex, collaborative, and iterative fashion in which policymaking around urban agriculture occurs. Civic activists influenced the scope, intent, and implementation of municipal policies; and vice versa, certain public officials felt that they were part of social movements, working in alliance with civic actors. At different moments in time and across different issue areas, particular sectors took the lead. In some cases, the issues were so entangled that there was no clear directionality or lead actor—the governance became truly hybrid. Civic-led coalitions created a social movement that brought the sense of a vibrant constituency to the awareness of local political leaders. At the same time, public officials helped organize conferences, engaged bureaucratic staff, and led the creation of visions and plans, all with the input of civic actors. These documents both triggered new programs and policy interventions and solidified narratives or ways of viewing the problems at hand. Thus, the artifact of the written plans, the policies they produced, and the ideologies they advanced all worked to bring food and agriculture into the municipal policymaking arena in new ways. While food was previously the "forgotten environmental issue," it is now becoming part of the policy discourse and of the municipal agenda.

7

CONSTRUCTING THE "GREENER, GREATER" CITY

Politics, Discourses, and Material Practices

Urban forestry and urban agriculture are constructed by networks of (human and nonhuman) actors, via overlapping and conflicting claims and through diverse material practices. We see these interwoven actor-networks unfolding at multiple physical sites, including farms, gardens, parks, street trees, and forest stands. Synthesizing across these two cases, this chapter returns to the three core themes of the book: the politics and governance, discourses, and materiality of urban nature. It also considers the way in which temporal change unfolds—with changes from above, below, and in-between.

Focusing on political power and the process of sustainability planning in New York City, we can identify networks, public-private partnerships, elite ties, and bureaucratic structures that were involved in goal creation, plan writing, implementation, reporting, research, and goal revision. This illustrates whose voices were heard and whose voices were ignored in sustainability planning. Yet politics did not stop once the plan was written; the practice of natural resource management is shot through with power. Sustainability planning is nested within relationships and histories that extend from before the plan was written and beyond its bounds.

Examining the construction of urban nature as a discursive process, there are key rhetorical claims associated with urban forestry and urban agriculture, particularly as positioned within the documents *PlaNYC, Food in the Public Interest, FoodNYC,* and *FoodWorks*. In both cases, advocates made claims about the benefits of these types of nature (trees, farms) to human health and well-being, neighborhood livability, the environment, and the economy. These claims reveal the values, ideology, and strategies of the developers of the plans, including the tacit understandings of sustainability that they embrace. Notions of growth, entrepreneurialism, and intercity competition were threaded throughout the rationales of the plans. Urban forestry and local food system narratives were crafted and burnished to support the competitive city discourse. At the same time, counternarratives related to justice and equity are featured in storylines as well. Yet certain claims and goals were bracketed out of the planning processes for failure to align with the dominant values and discourses of the plan—the principle of economic growth was never challenged.

Growing, planting, pruning, and watering trees, creating and maintaining gardens, and growing crops are materially and spatially differentiated processes. Further, nonhuman actors, both biotic and abiotic, play a role in the policymaking and natural resource management story. Changes to urban socio-nature are not occurring on a blank slate; they are being inserted into the already-existing, historically accreted built form of New York City. This urban spatial form constrains sustainability policymaking and natural resource management. In particular, as of 2007, New York City was highly developed, with few vacant parcels not already programmed for future use. This spatiality was made all the more complex by the numerous authorities with property jurisdiction on different pieces of urban space. This sense of a "lack of space," overwhelming density, and jurisdictional complexity shaped narratives, political tactics, and material practices related to urban forestry and agriculture. Finally, trees and farms have innate capacities that allow humans to make claims, extract value, measure and monitor changes, and find meanings in support of their policy and planning objectives. Conversely, other characteristics of trees and farms hinder our ability to make those claims. These capacities and limitations influence the way in which the nonhuman actants are represented and treated in sustainability plans.

Politics and Governance

Within the urban environmental policy arena, there are a number of commonalities between the actors, social structures, and relationships involved in forestry and agriculture. In both cases, the mayor played a direct role as a public official authorized to spend municipal funds, appoint agency heads, and direct strategic initiatives. *This* mayor (Bloomberg), in particular, played a role as a private philanthropist who could leverage personal finances and networks in support of his signature efforts. But the Office of the Mayor also played an important role, as potential mayoral candidates vied for public recognition and a policy platform with which to make their name. In moving from planning for to implementation of natural resource management, civic groups and the public had opportunities in both cases to be engaged, particularly via public-private partnerships and volunteer stewardship programs.

At the same time, urban forestry and urban agriculture are treated differently in the political maneuvers of public, private, and civic actors in New York City. Urban forestry and sustainability planning are largely government-led, institutionalized processes. Civil society groups were invited into the policymaking and implementation of both PlaNYC and MillionTreesNYC in a controlled, formal manner such as in public comment periods, consultations, and advisory boards. Only certain, professionalized nonprofit groups with access to key resources—such as elite ties and substantial fundraising capacity—were able to trump this process and insert themselves in a central role. Conversely, urban agriculture functions more like a civic-led social movement with key state and private sector allies, where diverse groups advance policy agendas and attempt to shape the policy discourse.

The Mayor and His Others: Agenda Setting

The presence or absence of a top-down endorsement from City Hall is one key difference between urban forestry and agriculture in New York City. The MillionTreesNYC campaign was developed as one of the 127 strategic mayoral initiatives of PlaNYC. In contrast, urban agriculture in New York City is better understood as part of a broad and emergent, primarily civic-powered social movement around food, sustainable urbanism, and community-managed open space that has historical roots in prior cycles of urban disinvestment and fiscal crisis. This is not to suggest that forestry lacked a network of concerned civic actors, nor to suggest that agriculture lacked crucial public allies and supporters. However, at this point in their development, there was a "center of gravity" within each case in terms of which sector (and which particular actors) drove the agenda.

Critiques have been leveled against PlaNYC for its top-down nature and the way in which it was so thoroughly a set of mayoral executive initiatives, repackaged together as a "plan," without the public involvement and legislative oversight required of an actual city plan (Angotti 2010). This book sheds light on the diversity of actions and initiatives that constitute planning and—in particular—the incredible power of mayoral-led strategic planning and policymaking. Bloomberg was well positioned with a municipal surplus at the time of writing the plan, recognized global leadership on climate adaptation and planning through the C40 network, and had a legislatively renegotiated third term of office, leading to twelve years in the position. The mayor, his deputy mayors, and the

City Hall staff maintained tight control over agenda-setting, question formation, scope delimitation, timelines, participation, framing, and the writing of PlaNYC as well as the heavily scripted rollout of the plan to the public and the media. The next ring of engagement involved key bureaucrats from city agencies making arguments, staking turf, and seeking to influence goalsetting through acts of interagency coordination and compromise. This is because PlaNYC was an undertaking of the executive branch rather than a series of local laws and resolutions that might emerge from a legislative initiative of the city council as was the case in the *FoodWorks* plan. Thus City Hall and city agencies were primarily responsible for the inclusion of forestry-related goals and the lack of agriculture-related goals in the first iteration of the plan. At the time, community gardening and urban agriculture were not part of the mayor's vision of a greener, greater New York City.

This account, however, is too cut and dried. In both cases, we see the role of networked actors, public and private, outside of this inner circle of decision-makers, in influencing the development, implementation, and—later—revision of PlaNYC. First, a Sustainability Advisory Board with organizations from civil society, business, and other branches of government was built into the PlaNYC process from its outset. Second, and perhaps most notably, the nonprofit New York Restoration Project (NYRP) was able to use the status and connections of its celebrity founder in order to gain access to the highest levels of power within New York City government. Driven by motivations completely outside of the PlaNYC process, NYRP's aims were then folded into PlaNYC, and the truly hybrid MillionTreesNYC was created. No such equivalent endeavor existed in the case of urban agriculture. The issue was discussed, but without a powerful champion, it did not gain traction. City Hall viewed the advocates for urban agriculture as outsiders to the PlaNYC process with little savvy on how policies work and programs are made. Gardening was considered by some as quaint or antiquated, and large-scale urban farming was considered unfeasible in the built environment of New York City.

Yet planning is not a moment-in-time endeavor. Particularly given the requirements for reporting, monitoring, and updating, we can think of the implementation stage as very much a part of the plan. It is in this stage that the forestry case opened up to a much broader set of actors, through its advisory committee and subcommittees; partners in stewardship, education, green jobs, and research; and its engagement with the public as volunteers and stewards. MillionTreesNYC—through its resources, public visibility, timeline, and sense of momentum—became the "train that left the station" for the already-existing advocates and allies of urban forestry. A reorganization of the network began to occur as new ties were created and institutionalized throughout the life of the campaign.

In response to both public critique and the changes in the economy, the 2011 update to the plan—or "PlaNYC 2.0"—was approached in a more inclusive manner. Out of conscience and necessity, more voices were brought into the process at earlier stages through both greater numbers of consultations with advocacy groups and more extensive public fora during the writing of the plan. The former were seen as a way to vet goals, but also to assess what resources outside entities might contribute toward PlaNYC ends—neoliberal policymaking during fiscally dire times. The latter were potentially a chance to gather ideas. But as the decision-makers felt that they had already considered most ideas, expanded public engagement was more often used as a means to create buy-in and feelings of inclusion to help build constituencies in support of the plan. At the same time, the process of revising the plan was a key advocacy moment for urban agriculture and community garden advocates who saw recognition within the plan as a politically meaningful step for the movement, even if it came without the substantial financial resources associated with the first edition of the plan.

The urban agriculture case also reveals the way in which other municipal actors outside of City Hall, working in concert with civil society coalitions, in turn influenced PlaNYC. While *FoodWorks, FoodNYC,* and *Food in the Public Interest* have policy implications in their own right, they also helped create pressure on City Hall and the Office of Long-Term Planning and Sustainability (OLTPS, now MOS) to incorporate food as a cross-cutting issue in the PlaNYC 2.0 update. Ideas and proposals around food and agriculture that primarily civil society groups had previously embraced came to be accepted by a handful of political figures, who readied the ground for at least tentative acceptance of these ideas by City Hall. In part, the impulse to incorporate food into PlaNYC was pure politics; Bloomberg and Quinn both were working to assert their policy turf in the environmental arena, and Bloomberg did not want to be seen to lag behind Quinn. Meanwhile, Quinn's team worked to ensure that *FoodWorks* would be released before PlaNYC 2.0 was. As a result, there is clear, substantive overlap in the recommendations put forth in the two documents, although the scope of *FoodWorks* is much broader. These incremental, overlapping policy inroads reveal the slow, tectonic shifting of municipal viewpoints toward food and agriculture.

Public-Private Partnerships and Coalitions: Implementation and Advocacy

Moving further into implementation, public-private partnerships are a key governance form in both cases, although they were certainly most central in the urban forestry case. The MillionTreesNYC campaign is a hybrid entity comprised of DPR and NYRP. Joint goals, a formalized memorandum of understanding, and a shared external-facing identity through the campaign's website, logo, branding, and messaging characterized this partnership. The participating organizations remained quite distinct, however, in terms of their staffs, core missions, office spaces, field sites, planting locations, and operational routines. Differences in expertise, jurisdictional authority, and the ability to fundraise and spend on this initiative were all core motivations behind the partnership. It is important to note that this partnership was forged in the economic boom year of 2007 and was, therefore, more of a product of routine, collaborative modes of governance than a response to crisis or austerity measures. Some scholars argue that public-private partnerships can arise out of financial necessity or austerity, as municipal governments look to fill gaps in their budgets. In this case, the partnership did leverage private resources, but the campaign reflected a net increase in public expenditures as well.

In the urban agriculture case, the formal public-private partnership supporting school gardens, Grow to Learn, was created in 2010 and modeled somewhat after MillionTreesNYC. It was comprised of partners from the Department of Education (DOE), DPR's GreenThumb, and the nonprofit GrowNYC, with funding from the Mayor's Fund and several private funders. However, the MillionTreesNYC campaign was more prominent and better funded than the Grow to Learn effort, as a result of their different scope, timelines, and origin stories. Nonetheless, Grow to Learn reveals the way in which institutional forms can perpetuate and become embedded in everyday practice the more they are used and seen as trusted or proven.

If the formalized public-private partnership is the central institutional arrangement in the forestry case, then complex and loose coalitions are the signature of the urban agriculture case. Coalitions involve an alliance of organizations and individuals working toward a common end or set of ends. These coalitions vary in their emphasis and composition. Some are catalyzed by formal gatherings and planning efforts led by public officials. Stringer helped to trigger coalitions through his leadership in organizing the "Politics of Food" conference at Columbia University and the "Food and Climate Summit" at New York University. These conferences and the documents they produced—while valuable in their own right—also helped to set the stage for the less broadly inclusive, but more formally institutionalized, *FoodWorks* plan led by Quinn. Other coalitions were fully civic-led efforts. For example, the Brooklyn Food Coalition (BFC) and the Food Systems Network of New York City (FSNNYC) brought together disparate actors to engage in mutual learning, public education, on-the-ground projects, and advocacy—albeit for the finite period that these coalitions lasted.

There are also coalitions that are historical legacies of prior efforts and crises that were reengaged and energized by current issues, as was the case with the New York City Community Garden Coalition (NYCCGC). The NYCCGC developed out of the community gardening crisis of the 1990s when then-Mayor Rudolph Giuliani tried to put up several hundred gardens for public auction. Since that acute moment of struggle, a broad network of participants in, advocates for, and allies of community gardens formed. This group remobilized in 2010 when the attorney general's memorandum of agreement that set the terms for garden management and dispensation expired, and DPR had to create new garden rules. NYCCGC continues to advocate on behalf of gardens that remain subject to development and displacement, including through legal action and lawsuits. Finally, still other coalitions draw specific attention to diversity and the potential imbalances or acts of exclusion within the food movement, as in the case of the Black Urban Growers.

There are other networks of individuals that might not be characterized as "coalitions" but are similarly groups of individuals connected through or served by a common umbrella organization. In many cases, these umbrella organizations are the broker or bridging groups described in chapter 1 (see also Connolly et al. 2013). These networks include participants in food- and agriculture-related programs such as Just Food, New York Botanical Garden (NYBG), and Brooklyn Botanic Garden (BBG), as well as gardeners interacting with Green Guerillas and gardeners working on sites owned and supported by NYRP, Trust for Public Land (TPL), and GreenThumb. Even if not all of these participants are deliberately motivated toward advocacy in a social movement vein, they are connected by loose ties that allow for sharing knowledge, ideas, information, and storylines.

Similar sets of organizational members and program participants exist in the urban forestry case, as members of the public interact with the forest-related programs of DPR, NYRP, Trees New York (TNY), and the MillionTreesNYC Stewardship Corps (including GreenThumb, BBG, QBG, NYBG, and the Greenbelt Conservancy). Yet because this arena was never threatened by an overt crisis like the one that affected the community gardening world, and because forestry does not involve the same deep history of reappropriation of land and community-based management that community gardening entails, these groups do not (yet) operate as a social movement. In acknowledgment of this prior lack of a movement, leaders within MillionTreesNYC had attempted to "build a social movement around urban forestry" (respondent 15). Toward that end, as the campaign progressed, MillionTreesNYC formalized its outreach and stewardship efforts into the TreeLC program, which used mini-grants and train-the-trainer models in order to help grow the pool of people touched by the campaign who would engage in tree care. Since the conclusion of the MillionTreesNYC campaign, those programs have further expanded and developed under the auspices of DPR, to include working with the public on street tree care and in natural area restoration work.

Given the political developments and changes in governance described over the course of just 2007–2011, what will future examinations of the social networks of urban forestry and urban agriculture (i.e., like the visualizations provided in chapter 1) find in the networked structure of these two arenas? Perhaps we will find more polycentricism and a broadening of civic actors involved in urban forestry as an urban forestry social movement takes hold. And perhaps we will find more centralized government engagement as urban agriculture becomes more embedded in municipal policies and programs. Perhaps we will see a greater overlap in the particular actors that are involved across both of these arenas, as a greater understanding of cities as social-ecological systems (not as fragmented silos) continues to develop. Time will tell if this sort of convergence—or if future divergence—occurs.

Discourse

Discourse and storylines have political and material effects. And vice versa, politics and materiality help constitute discourses. Language has the power to assert ideologies and to enforce (and reinforce) hegemonic or dominant thinking; alternatively, it can be deployed critically to challenge the existing social order. The narrative framing of urban forestry and agriculture in sustainability plans, green infrastructure campaigns, and food systems plans can be examined as enactments of networked environmental governance.[1] In so doing, I find a number of key themes, each of which are presented here. First, *quantification* plays a crucial role in the construction of rationales supporting the investment in urban nature. Second, the language of the *entrepreneurial city* is prominent in these plans, focusing on competitiveness, growth, and efficiency. Third, there are different *conceptions of justice* that are present in PlaNYC, MillionTreesNYC, and the food plans. Fourth, there are *shifting narratives of food*—in terms of systems, scale, and scope—in the policy arena. Finally, I uncover *unstated assumptions about the role of the state and the public* that permeate the plans in their taken-for-granted absence.

Quantification and Counting

Advocates for urban forestry and agriculture make claims about the benefits of both of these types of nature, leveraging research data, quantitative arguments, emotional appeals, and storytelling to make their case. Many respondents discussed the particular importance of quantitative evidence in making a compelling argument to the Bloomberg administration, which sought to apply business-like reasoning to municipal governance. PlaNYC, in particular, emphasized the setting of numeric targets that could be tracked via metrics on a quarterly and annual basis. In the case of urban forestry, bureaucrats seeking to enhance their street tree planting and forest restoration work marshaled decades of city data on trees and parklands. As well, an institutional infrastructure—the work and research on urban forestry of the U.S. Forest Service and numerous university departments—aided the DPR.

While City Hall received a range of characterizations of the urban forest, the monetized view of ecosystem services, based off of the STRATUM and iTree models, convinced decision-makers of the service provided by street trees. These models allowed bureaucrats to argue that trees were a "sound business investment" that would provide more financial returns than they cost to install and maintain and that the investment would mature as trees grew. In particular, the substantial gains in real estate value projected in association with street tree plantings struck home with the decision-makers. Although somewhat tacit in the language of the plan, these potential gains fit with a view of livability that would attract and retain certain sorts of (high-income) residents. PlaNYC states:

> There is no formula for the perfect New York City block. But neighborhoods with trees are generally more pleasant and beautiful than those without; sidewalks that encourage walking, with room for strollers, and gawkers, and *go-getters*, are more interesting and enjoyable than narrow strips of concrete. Our plan for open space will help bring to life the unique beauty of each of our neighborhoods. . . . It means filling out the remaining barren streets with trees that will add shade, color, cleaner air and *higher property values;* and it means encouraging an active, vibrant public realm as essential to the life of our city. (City of New York 2007, 36–37, emphasis added)

While the real estate returns may have convinced City Hall, the multiple benefits of trees provided nearly limitless rationales for investing substantially in the urban forest. In the plan itself, the investments were framed primarily as contributing to air quality and enhancing open space. But arguments were also made throughout the MillionTreesNYC campaign about the diverse benefits of trees, including public health gains, stress reduction, and educational benefits. Even when new research or data contested some of these relationships—as in the association between trees and asthma reduction—decision-makers retained an overall feeling that trees are good for the environment, the economy, and society and that they make sense as a municipal investment.

Counting practices were crucial to the implementation of the MillionTreesNYC campaign. Weekly, monthly, quarterly, and annual reports summarized the number of trees planted by land jurisdiction and site type, while a large red LED clock (located at City Hall, DPR, and the offices of thirteen other city agencies) counted down the number of days left in the mayor's term (Barbaro 2008). There was a real sense that staying on track with the time-delimited program necessitated planting large numbers of trees ahead of schedule. Indeed, in late 2012, the timeline of the campaign was adjusted from ending in 2017 to ending in 2015. Counting individual trees also served as a way to engage the public over the course of the campaign. The inaugural tree planted was marked with a sign saying "One in a Million," and the 500,000th tree planted was a cause for celebration and media fanfare.

Several of the critiques of urban agriculture from decision-makers related to the relative lack of data about urban farming. This is largely due to the fact that many of the benefits and services of urban agriculture were previously unquantified, particularly in terms of the social benefits of farms and community gardens. Historically, these sites have been largely ignored or overlooked by research-supporting entities, with the exception of extension organizations. Many advocates noted that the USDA's Census of Agriculture (http://www.agcensus.usda.gov/) defines farms in such a way that it excludes most urban sites. Moreover, urban agriculture and community gardening have deep ties to activists and grassroots organizations, which have in the past focused more on action than on research. Though DPR's GreenThumb does a substantial amount of work to register and track gardens over time, as community-managed open spaces rather than bureaucratically controlled sites, the diversity of land use and natural resource management practices on the hundreds of gardens citywide presents challenges for data gathering and tracking.

This lack of data is clearly changing, as evidenced by the proliferation of academic research and citizen science on urban agriculture, followed by new municipal efforts at tracking and counting. Of particular note in New York City are the Columbia Urban Design Lab's *Report on the Potential for Urban Agriculture* and research on the New York City foodshed; the *Five Borough Farm* project of the Design Trust for Public Space; and citizen science efforts to understand the impact of urban agriculture, like Farming Concrete and Seeing Green. These efforts varied in their research questions and methods employed, but all attempted to develop and apply metrics and gather data to help understand the current impact of and potential for further development of urban (and regional) agriculture. In response to the various civic-led initiatives and the food-related plans of the Manhattan borough president and the city council. City Hall and the municipal agencies responded by placing a greater emphasis on quantification in relation to agriculture and gardening. They created an Urban Agriculture Taskforce to build a publicly accessible, online database of city-owned vacant property. This taskforce involved the participation of a wide range of agencies beyond DPR, as land can be held in many different jurisdictions and requires careful accounting and participation of all relevant landholding agencies.

Competition, Growth, and Efficiency: The Language of the Plan

PlaNYC's subtitle was "A Greener, Greater New York City," which positioned the document right from the start within the competitive city discourse. This entrepreneurial view of intercity competition was embedded deeply in the DNA of the plan, as well as in its rhetorical presentation. According to the introduction to the plan, New York was competing with U.S. cities and other global cities for residents, businesses, and tourists: "Our competition today is no longer only cities like Chicago and Los Angeles—it's also London and Shanghai. Cities around the world are pushing themselves to become more convenient and enjoyable, without sacrificing excitement or energy. In order to compete in the 21st century economy, we must not only keep up with the innovations of others, but surpass them" (City of New York 2007, 10). Public officials sought to enable successful competition through investments in the urban realm. Trees were believed to have multiple environmental, public health, and economic benefits that would enhance neighborhoods and help New York City compete in attracting discerning residents. For example, tree planting was presented at several points in the plan as contributing toward New York City's goal of having "the cleanest air of any big city in America" (City of New York 2007, 11). This sense of comparison and competition was also embedded in the graphics presented in the plan. A map of New York City's urban tree canopy levels by neighborhood was presented alongside a bar graph showing how New York City's citywide tree canopy level compares to that of Atlanta, Austin, Washington, D.C., Boston, Seattle, Baltimore, Wilmington, Philadelphia, Jersey City, Milwaukee, and Chicago.

In contrast, urban agriculture, with its historic association with periods of housing vacancy and decline followed by civic investment of sweat equity, did not fit as cleanly with this narrative of urban competitiveness, at least at the time of the plan's writing in 2005–2006. In New York City, gardens are generally located in the neighborhoods that saw the greatest decline in the 1970s and 1980s in terms of the flight of capital and (particularly white) residents, such as the South

Bronx, East Harlem, the Lower East Side, and Central Brooklyn. Some decision-makers saw community gardening and urban farming as antiquated, marginal, or idealistic practices left over from another era. Only once gardening came to be reframed as urban agriculture, associated with rooftop farming, greenhouses, hydroponics, and local entrepreneurialism in the late 2000s, did decision-makers see a competitive value in incorporating it—however nominally—into PlaNYC 2.0. *FoodWorks* also illustrated this reframing as it claimed that strengthening urban and regional agriculture could "inspire urban agriculture innovation" and enhance the competitiveness of the New York City region, including its competitive advantage in certain agricultural products (NYC Council 2010).

PlaNYC placed great value on growth and efficiency. The starting question that Deputy Mayor Doctoroff posed to his City Hall staff and agency representatives was: "Given an increase in 1 million new residents in New York City by 2030, how can municipal resources be used most efficiently and effectively to serve those residents?" A policymaker elaborated on this process of planning for efficiency in a growing city:

> We started looking at the different ways that we could plan for growth citywide, as opposed to neighborhood-by-neighborhood. And the conversation was always starting out with: "What more resources do you need? So if this neighborhood grew by x number of people, how many more school seats would you need. How many more firehouses? How many more waste transfer stations?" Those type of municipal uses. And as the conversation evolved, you got the same answer back, which is, "we could use more physical resources." Like you could use x number more megawatts of energy per person. Or you could tell people to use less energy. And then more people could use the same amount of resources that we have now. . . . There's a physical answer, which would mean more land, more money, more whatever; or there's the sustainability answer. So that's how PlaNYC evolved. (respondent 26)

The plan discussed both the efficiency *of* cities (because of their density, and as opposed to other patterns of human habitation) and the need for efficiency *in* cities (in terms of the use of resources, energy, and land). Efficiency was thus a flexible concept that asserted the importance of and need for investments in New York City because it offered an environmentally friendly model of living. As PlaNYC stated, "And we will also make a difference in the fight against global warming simply by making our city stronger: By absorbing 900,000 new residents—instead of having them live elsewhere in the United States—we can prevent an additional 15.6 million metric tons of greenhouse gases from being released into the atmosphere" (City of New York 2007, 13). Efficiency rhetoric highlighted the choices that had to be made in planning and development without ever challenging the first premise of growth.

Differing Conceptions of Justice: Distributional, Procedural, and Social

While competitive city and growth-centric rhetoric abounded in PlaNYC, there was also a less prominent, but still important, justice narrative. Perhaps most notably, the goal that every New Yorker should live within a ten-minute walk of a park reflected a sense of distributional justice that has proliferated within planning circles, particularly those focused on urban greening and livable cities. This goal was positioned on the front page of the open space chapter as the overarching aim to which all of the initiatives would contribute. A visually compelling map showed all the areas in New York City that were projected by 2030 to be within a ten-minute walk of a park, if all of the PlaNYC initiatives were to be successfully implemented. The ten-minute walk concept was developed *after* various divisions within DPR proposed and vetted initiatives; this argument provided a conceptual rationale for explaining the many disparate DPR initiatives to the public. In terms of urban forestry, distributional unevenness was highlighted through a map in the plan that showed the street tree stocking levels and how they varied by neighborhood. This metric, along with considerations of air quality and asthma incidence, was used to select the Trees for Public Health (TPH) neighborhoods where intensive block planting was focused first. DPR bureaucrats sought a publicly defensible rationale for correcting unevenness in the urban forest that had manifested due to an individual request-based system (i.e., "the squeaky wheel") for planting street trees. Similarly, NYRP was interested in prioritizing planting in communities of need and focused its private tree plantings and giveaways in TPH neighborhoods first.

Despite the clear and laudable attention to distributional justice, PlaNYC and MillionTreesNYC lacked an attention to procedural justice or devolution of power. (The procedural issues related to PlaNYC's conceptualization and goal setting have already been discussed above.) In the case of MillionTreesNYC, for the most part, the planting of trees was the result of professionalized processes managed by the city, in partnership with one nonprofit group, and implemented through private-sector contracts with the landscaping industry. While the campaign sought vigorous volunteer engagement and participation in the planting of forest restoration sites and the maintenance of street trees, it was not wholly rooted in a community forestry approach.[2] Instead, it used volunteer labor to help ensure the survival of street trees beyond their two-year contractual guarantees and to reduce the cost of large-scale reforestation plantings. Yet DPR hoped that volunteers would come to feel more invested in reforestation sites and street trees on their blocks in the long term, and stewardship programs continued to evolve over the course of the campaign and beyond. MillionTreesNYC worked to create a comprehensive volunteer program, with tree planting events serving as one of the key points of contact between the public and the campaign.

The food plans were more overt in their commitment to social justice, the poor, and the underserved than was PlaNYC. In both Stringer documents, the problems of hunger, obesity, and diet-related diseases were framed as problems prevalent among low-income populations. So, too, did *FoodWorks* note that hunger is due to poverty and inequality—not simply to an inadequate food supply. Moreover, these reports, which were released in 2009 and 2010, acknowledged the context of the economic recession and the impact it has had on people's lives. And the reports made policy recommendations to address food insecurity and low-income "food deserts." In addition to supporting the philosophy that the government should help address these issues, the reports called for a broad-based social movement to effect change (Stringer 2009). The food system was also discussed as an economic sector and as a potential source of local jobs; calls for fair wages in the food sector showed alliances with labor movements. Finally, the Stringer documents and *FoodWorks* called for a formal role for citizens in policymaking via a proposed Food Policy Council, which is a form of collaborative, stakeholder-driven governance not currently in existence in New York City.

Changing Narratives of Food: Systems, Scale, and Scope

The relative absence of food from PlaNYC was countered by its proliferation in networks that formed external to—and exerted influence on—the plan. Chapter 5 explored the numerous narrative themes and material practices that emerged surrounding food and agriculture in urban and peri-urban areas surrounding New York City. The broadly inclusive conceptual framework of a "food system" was used to weave together these disparate threads of interest into one coherent (if complex) discussion around how food is produced, processed, and consumed. As described in chapter 6, these diverse networks of activists and public officials coalesced to produce planning documents, which in turn further solidified the New York City food system narrative. Much of the work that civil society activists did consisted of bringing food into the arena of environmental issues that were on the table for municipal policy discussion. This focus on the food system was melded with a definition of sustainability that focused on economic growth, environmental quality, and public health.

Drawing attention to the scalar politics of food through New York City's connection to a one hundred– or two hundred–mile foodshed (or even all of New

York State) was an important approach of these plans. Although urban agriculture is a part of this food system, it was certainly not the most prominent or visible aspect within the policy documents. Much attention was paid to rural producers throughout the region and state and to how to build urban-rural linkages. There was also an emphasis placed on consumption and processing because of the links to local job creation, the importance of the food retail and restaurant sector, and the role of the city as a purchaser of foodstuffs. Further, since New York City's local food system is fundamentally enmeshed in a globalized food system, the plans acknowledged the limitations of actions that occur locally—but pursued them nonetheless. Thus, the plans occupy a unique rhetorical space that represents the complex, multiscalar, historic, and embedded nature of the problems they are trying to address, but then the plans pivot to propose pragmatic, first-step solutions that are limited in scope and implementable locally.

Beyond the bounds of the plans, we see calls for more radical action and change, particularly in the food justice and food sovereignty movements. Interest in shifting power to the poor and disenfranchised, forming transnational alliances with rural peasant producers, and creating solidarity with a wide range of progressive and radical social movements are all discussed and pursued. Because food politics are scalar, these alliances between regional and small-scale urban and rural producers critique and challenge the global, corporate food system. Localization and regionalization is—in many cases—a direct counter to the perceived flaws of the globalized, industrialized food system. The localization narrative, though, is coupled with transnational alliance building that transcends place, a frequently employed social movement tactic (see, e.g., Keck and Sikkink 1998).

The local food movement in New York City—both within and beyond the plans—placed particular attention in its rhetoric and policies on youth, in part because of the epidemic of childhood obesity. The sense of crisis and urgency associated with childhood obesity and diabetes built awareness of the need to focus on access to healthy food; prominent public figures like Michele Obama and funders like the Robert Wood Johnson Foundation have advanced discussion of these issues. The emphasis on youth also allows for presenting convincing normative claims about need, justice, and fairness as related to healthy food access. America has a long tradition of social services and welfare focused on the needs of the "deserving poor," which has often included children. For example, one of the longest-lasting federal safety nets is the Women, Infants, and Children (WIC) supplemental nutrition program. Some local charitable efforts to remedy child hunger, malnutrition, and obesity fit in this lineage. At the same time, more radical efforts focused on shifting power and agency in the food system also work to empower youth as leaders, such as the Flip the Table Youth Food Policy

Council. Finally, focusing on youth can also meet very real needs by leveraging the power of the massive New York City public school system as a purchaser and preparer of school lunches and breakfasts. Thus, for varying strategic and ideological reasons, youth are an important focus in much of this work.

Finally, we can unpack the implications of the current shift in language from *community gardening* to *urban agriculture* common in many policy circles.[3] Some activists contest the shift and draw attention to the fact that nothing is so *new* about urban agriculture; it is part of a long lineage of growing food in cities, and people have been doing it for decades with little fanfare in the press and policymaking circles. The new attention to urban agriculture raises all sorts of questions about inclusion and institutional racism—why now is urban agriculture so appealing? And might it have to do with the demographic profile of some of the current wave of participants in the practice (young, white, educated) as compared to gardeners and farmers from low-income communities of color? Other activists contest the foregrounding of food production over the other important functions of community gardens. Many of these garden sites were created to promote neighborhood stabilization, to serve as gathering spaces—as recreational, artistic, political, and intergenerational spaces. While the hundreds of community gardens in New York City are highly variable in their management practices, in many cases the growing of plants and crops is more of a means than an end. By casting community gardens whole cloth as part of urban agriculture, there is a danger that food production will eclipse the many other important reasons why we might want gardens (or even farms!) in the city. In fact, many of the current practitioners of larger-scale urban food production recognize that one of the most valuable contributions of these sites is to educate urban residents about food, agriculture, and ecology. These sites are inherently multifunctional and are about much more than just food. Therefore, it is important that the current attention to urban agriculture not eclipse the history of gardening in the city that long precedes the current wave of interest in hyperlocal food.

Key Assumptions of the Plans: What Role for the State? What Role for Citizens?

When conducting discourse analysis, one of the most important—but also most challenging—tasks is to be mindful not only of what is said, but also of what is *not* said. The tacit assumptions that underlie a plan, particularly with regard to the appropriate role for the state, the economy, and citizens, can often be so dominant, hegemonic, or taken for granted that they are not explicitly stated. PlaNYC represented a unique form of Bloomberg-era "roll out" neoliberalism (see Peck and Tickell 2002). The state expended resources in order to make the city more hospitable (and competitive) to businesses and residents. Moreover, the focus of this plan was *not* primarily on social justice or the redistribution of wealth. The social components of sustainability were couched more in a language of public health than of justice, which provided flexibility that allowed for neighborhood investments without radical social change. Nor did PlaNYC focus on creating a suite of new environmental regulations. When the city did attempt to create regulations, as was the case with the proposed congestion pricing for driving in lower Manhattan, public opposition and a state-level veto thwarted this effort. In the 2011 update to the plan, the banning and phasing out of #6 and #4 home heating oil was similarly controversial and opposed by many building operators and developers.

PlaNYC's decision-makers focused on goals and initiatives that were "actionable" or "implementable," thereby limiting the scope to efforts that were politically and financially feasible. Very little was proposed that prominent political forces would oppose, and nothing was articulated that the city (along with private partnerships, tristate cooperation in the metropolitan region, and selected state support) could not fund. For example, while there was a national conversation about the importance of "green jobs" to the sustainability of cities, the plan did not propose large-scale government funding to train or employ the unemployed in these fields. This was not considered the role of this municipal plan. When green jobs training and employment programs were created, they were small in scale, limited in duration, and supported with private funding. Other potential initiatives disregarded as "future ideas" included technologically dependent efforts, like vertical farms or large-scale hydroponic agriculture. Although the technology does exist for these approaches, it is not yet affordable, nor is the private market demanding it on a broad scale, so it is not seen as the role of government to intervene. Most strikingly, the notion of doing only what is implementable produced the rationalizations of PlaNYC 2.0's goal setting, which merely articulated what agencies were *already* doing anyway, with no new commitment of human or financial resources.

Another one of the main assumptions of PlaNYC and MillionTreesNYC was that trees, parks, and the public right of way (PROW) should be considered as green infrastructure provided by the state for the benefit of the public. DPR officials ensured that City Hall specified that residents could not refuse street trees, just as they could not refuse a stop sign, a fire hydrant, or any other piece of infrastructure in the PROW. This was in anticipation of potential complaints from some homeowners who view the sidewalk in front of their house as an extension of their private property, although in actuality it is a public space. The PROW is something of a jurisdictional "grey zone" because residents are required to remove garbage, debris, and snow from the sidewalk and maintain the sidewalk in good condition but are prohibited from cutting roots or pruning trees without the city's permission or a citizen-pruner certification (Rae et al. 2010; NYC DOT 2013). At the same time, might this notion of green infrastructure be at odds with the cultivation of community engagement, which the MillionTreesNYC campaign tried to do through its various marketing, stewardship, and education efforts? We do not ask residents to maintain roads, tunnels, bridges, or the utility grid; "infrastructure" is generally understood to be a public good.

The food plans differed substantively from PlaNYC in how they bounded the role of the municipality specifically and role of government more broadly. Both the Stringer documents and *FoodWorks* articulated principles and goals that would require substantial collaboration and investment at the state and federal levels. For example, acknowledging the importance of the public food safety net, the reports called for changes to federal benefits programs and the Farm Bill. Still, these plans did not use a regulatory approach. Although the city council is generally known for being more progressive than the mayor's office within New York City government, *FoodWorks*'s policies related to urban agriculture operated primarily through rhetoric, programmatic support, small tweaks to bureaucratic practices and reporting, and rather innocuous local laws. Similarly, in the Stringer reports, government was viewed as a convener, a purchaser of foodstuffs (with implications for demand-side policies), and a partner with the private sector—but its job was not to mandate. In discussing the food system, the plans repeatedly articulated the important role of the private sector in all stages of production, processing, distribution, consumption, and postconsumption. Critiques of the private sector's role in the creation of food insecurity and food injustice were more overt in the Stringer documents, but all reports noted the need for the public sector to work hand in hand with the private sector to achieve changes to the food system.

Taken together, these themes present us with a clearer picture of how PlaNYC was ideologically positioned. Quantification, competition, growth, and efficiency were central to this sustainability plan. A focus on being actionable or implementable delimited the possibilities for more radical changes under the auspices of sustainability. At the same time, actionability placed emphasis on *doing* rather than just *planning* or envisioning—including through the tangible commitment of financial and human resources. We can trace the way in which these narratives are repeated and reified, or challenged and contested, throughout the broader fields of practice in urban forestry and agriculture. For example, some dimensions of justice were taken into account more than others, with strengthened attention to justice in the implementation of programs. Framing food as a commodity, a right, or a cultural product, and whether and how food can or should be grown in the city, remains an active area of debate among activists and policymakers alike. While narrative frames in planning documents can certainly be contested and can shift, we also can detect their traces and signatures in the way in which policies shaping urban space get enacted.

Spatiality and Materiality

Urban nature is not just the substrate for or outcome of human actions. Nonhuman actors play a role in local sustainability policymaking and green infrastructure campaigns as well. These plans are formulated in the context of the developed city, with a distinct spatial politics that flows from the condition of—or at least a perception of—"lack of space." The urban landscape is not an undifferentiated mass; rather, it is subdivided and crisscrossed into territorial domains. In particular, property ownership and land jurisdiction are key institutions that define and delimit natural resource management. At the same time, trees have needs, capacities, and limitations that shape the claims that are made about them and the practices that are used to manage them. Farms and gardens are complex assemblages of space, plants, soil, buildings, labor, water, air, and sun. In their sheer diversity of form and management, they present challenges for municipal policymakers' bureaucratic processes and rationalized views of nature.

Spatial Politics in the Developed City

Given a discourse of growth and interurban competition, along with policies that sought to balance development, open space, and infrastructural needs, PlaNYC engaged in the spatial politics of urban land use. Throughout PlaNYC—as well as the three food plans—there was an acute awareness of New York City as a developed city that lacks available space for large-scale projects (be they housing, commerce, open space, or transit):

> As virtually every part of our city grows, one piece remains fixed: the supply of land. That's why we must use our space more efficiently to accommodate growth while preserving—and enhancing—the city's quality of life. We must provide enough housing; but we must not allow the production of units to eclipse other neighborhood needs—the balance of open space, parks, retail, and aesthetics that is essential to a healthy community. With competing needs and limited land, we must unlock unrealized housing capacity, complete unfinished parks, and direct growth toward transit centers. By being smarter about our

> land-use strategies, we can realize the promise of an expanding population, while avoiding the pitfalls of unplanned and unbalanced growth. (City of New York 2007, 14)

Sustainability planning in New York City is not the same as sustainability planning in smaller, stable, or shrinking cities; it focuses primarily on small tweaks to the urban fabric that yield gains and efficiencies. One bureaucrat noted, "This is not Las Vegas," saying that planning in New York City cannot start from scratch or sprawl into the surrounding hinterlands; it must accommodate the densely built urban environment (respondent 49). Efficiency strategies proposed in PlaNYC include colocating facilities, extending hours of use (e.g., opening schoolyards during the summer, adding lights to parks), adaptively reusing underutilized or contaminated lands (including a chapter on brownfields), and enhancing the PROW. As part of the housing chapter of PlaNYC, the city engaged in a comprehensive effort to identify and bundle sites for the future development of market-rate and affordable housing, particularly emphasizing the creation of density near transit hubs in the outer boroughs. Active debates circulate about the appropriate balance of development and open space in the city, including whether precisely the efficiencies and density that makes New York City more environmentally "sustainable" are threatened by the creation of more open space (Light 2003; Owen 2009). Overall, population growth, an aging infrastructure, finite land, and climate change combine to create the planning quandary that drives PlaNYC. And the challenge of maintaining affordability amidst that population growth is acknowledged in the plan, even though the word "gentrification" is avoided.

From a spatial perspective, we can easily understand why growth-oriented municipal decision-makers would be more likely to embrace and fund urban forestry endeavors than urban agriculture. Street trees could be inserted into the already-existing urban fabric of the PROW. Thus, the trees enhance, rather than compete with, commercial and residential buildings by increasing real estate value. At the same time, DPR bureaucrats made equity arguments that every New York City resident deserved a high-quality public sphere, and enhancing sidewalks with street trees was one relatively affordable and rather innocuous way to do this. In contrast, agriculture generally requires open *space*. While there are certainly technological fixes including greenhouses and hydroponics, for the most part, crops require more contiguous land area than individual trees. Indeed, "lack of space" was seen by activists and policymakers alike as the greatest barrier to urban agriculture in New York City. *FoodWorks* identified these challenges: "New York City is the largest and densest metropolitan area in the country. Yet residents and organizations throughout the city have discovered creative ways to produce food within this environment using rooftops, vacant parcels and raised beds. Some new technologies are even able to grow food inside buildings. However, gardeners still face challenges protecting the gardens they have from development, finding new space for gardens, and navigating the city's approval processes" (NYC Council 2010, 26–27). Given this material reality and narrative framing, it is unsurprising that rooftop farming came to be championed as a "win-win" opportunity for advancing urban agriculture in New York City and received several mentions in PlaNYC 2.0. Underutilized roofs could be put to new uses, and rooftop agriculture does not compete with any other potential tax ratable in the way that a ground-level garden or farm does. Rooftop agriculture requires structurally sound, wide, and flat roofs and thus is well-suited as an adaptive reuse for the postindustrial city. Large warehouses and former factory buildings that were designed to support the weight of heavy machinery—such as those at the Brooklyn Navy Yard, site of the Brooklyn Grange Rooftop Farm—can safely be retrofitted to handle the weight of soil media and vegetation. There are just a handful of these rooftop agriculture sites at this time, but they have immense symbolic and rhetorical value as they represent a "way out" of the bind of the developed city.

Property Jurisdiction

From a governance perspective, jurisdictional authority over space plays a large role in shaping both sustainability goals and natural resource management practices more generally. Long before the creation of PlaNYC, DPR had been responsible for planting and managing trees in the PROW and in all city parkland. Thus, the agency (including both the CFH and NRG divisions) viewed these sites as part of its turf and mandate; the landscapes are monitored and tracked as part of everyday operations. The engagement of NYRP meant that the MillionTreesNYC campaign could span both public and private lands. In contrast, although DPR has a formal program supporting community gardens (GreenThumb), property jurisdiction of garden and farm sites is less stable, given the long history of resident reappropriation of vacant space, the garden crisis of the 1990s, and the subsequent garden settlement. Community gardens are also relatively small in terms of acreage, comprising approximately eighty-six acres citywide (Ackerman 2011, 33). During the development of PlaNYC, CFH was better positioned than GreenThumb to make its pitch to City Hall for investment in its programs. Tree planting seemed to City Hall like an "easy win" that could be rolled out under current DPR authority, enhanced with the private support of NYRP. In contrast, community gardening involves a whole host of nongovernmental actors, including formalized nonprofits and hundreds of grassroots community groups serving as local land managers.

Despite DPR's authority to plant trees in the PROW, the ambiguity over who "owns" the sidewalk produced challenges in the implementation stage of the MillionTreesNYC campaign. This ambiguity occurs, in part, because the city plants trees, installs signs, and builds and repairs the sidewalk, but residents are responsible for snow, garbage, and leaf litter removal. If residents do not maintain sidewalks in good condition, the city can repair the sidewalk but can bill the building owner for the cost of the work (NYC DOT 2013). Further, while the city and its private contractors are responsible for tree planting, with contract guarantees for the first two years of the tree's life, DPR hopes that New York City residents will engage in the care of this green infrastructure over its lifespan. Although MillionTreesNYC presented itself as a campaign to plant and care for a million new trees in New York City, the way in which capital monies were allocated and the processes through which DPR plants trees left a need for support in the long-term maintenance of trees that was widely acknowledged, even by the leadership of the campaign. Numerous programmatic efforts attempted to cultivate that stewardship, but it was not initially built into the DNA of this tree *planting* campaign as deeply as the rigorous attention devoted to the number of stems planted. An enhanced emphasis on stewardship programs did, however, develop over the course of the campaign and after its completion. In that way, we can see that the scale and import of the campaign fundamentally shaped the way in which DPR valued and staffed its stewardship programs going forward.

By wrapping afforestation and reforestation efforts under the PlaNYC tree planting mantle, DPR saw an opportunity to garner much-needed resources to manage "natural areas." These are sites managed for ecological functions, including woodlands, grasslands, and wetlands that are under the purview of DPR's NRG division. They are generally highly disturbed, landfilled, or vestigial sites, sometimes adjacent to roadways and other infrastructure. In addition, the "natural areas" generally have fewer resources and are less visible than recreational parks. Using PlaNYC funding, NRG attempted to create native, multistoried forests wherever the department deemed it ecologically appropriate. This involved aggressive and sustained management practices to remove invasive species physically (by hand removal, cutting, mowing), mechanically (by heavy machinery such as excavators), and chemically (via herbicide application), followed by the dense planting of native canopy trees—in an attempt to close the canopy and outcompete other species. DPR and its academic partners are actively researching whether these management practices will succeed in their aims of creating healthy forests that last over time.

NYRP's interest in tree planting was initially more conceptual than material; they lacked jurisdictional authority over urban trees, with the exception of a handful of sites where they served as the lead stewardship group. The founder, Bette Midler, was inspired by the *idea* of a million trees campaign and negotiated terms of a partnership at the highest levels of municipal government. Although NYRP was involved in park maintenance in a few Northern Manhattan parks and owned several dozen community gardens, the organization had not previously done large-scale tree giveaways, planting, or stewardship. Thus, planting site selection was truly a novel and evolving process. With DPR handling the planting on public lands, NYRP began first with identifying the largest private landholders in the city. NYRP began a partnership with NYCHA—the public housing agency—which operates somewhat outside of traditional executive agency politics because of its funding structure, governance, and unique history. After this relationship soured for a host of political and personality reasons, NYRP continued to work with other large landholders like the CUNY system and Co-Op City in the Bronx. Due to the high cost of these direct-planting efforts, NYRP later pivoted to focus more resources and attention on tree giveaways to individual residents for planting in their private lawns.

Community gardens are unevenly distributed across space throughout the city and are overseen by a patchwork of land managers. Sites are predominantly located in neighborhoods that experienced previous cycles of economic disinvestment and capital flight that led to the proliferation of vacant lots, most recently in the 1970s fiscal crisis. Despite ongoing development and gentrification, the more than five hundred gardens citywide are most heavily clustered in the East Village/Lower East Side and Harlem in Manhattan; the South Bronx; and central Brooklyn, including Bedford Stuyvesant and East New York. There are many fewer gardens in the lower density, more suburbanized areas of much of Queens and Staten Island, where private lawns are more common. Historically viewed as a temporary solution to the abundance of vacant land, gardening was a strategy individual residents and community groups used to reappropriate space for neighborhood use. Over time, these groups became more established, undertook lasting management of the landscape, experienced leadership transitions, and asserted their right to land tenure. When Mayor Giuliani tried to put up more than one hundred gardens for auction, the question of community garden land preservation came to the fore. The subsequent legal settlement divided the parcels between DPR-owned sites that would be preserved as gardens, other city-owned sites that could be subject to development, and privately owned lots bought by NYRP and TPL for protection as community gardens held in land trusts. To this day, gardens are characterized by their grassroots, bottom-up governance, and the community garden network is spatially dispersed and jurisdictionally fragmented, requiring different styles of communication and management than the rest of the DPR bureaucracy. Any efforts to try and change, manage, enhance, or challenge gardens necessarily works with (or faces opposition from) this complex, polycentric network.

Trees as Actants

Trees are also actants in this story. They have properties that shape and delimit what discourses are plausible and what management practices are possible. Their inherent, material characteristics provide a basis from which people with certain objectives (e.g., a "livable city") can mobilize claims and develop policies. And their biological needs for certain environmental conditions and resources that comprise their niche create challenges that managers meet through trial and error, expertise, rules, institutions, and relationships.

Although trees can be considered as parts of communities—such as stands or forests, or in aggregate measures like percent of tree canopy per unit area—the MillionTreesNYC campaign homed in on individual stems as the entities to manipulate in this effort. Trees, unlike rhizomes such as grasses, do grow in large, discrete units that can be understood individually. The very *treeness* of a tree is defined by its largely aboveground, single-stemmed form, which can also be contrasted to multistemmed shrubs. These individual units allowed for easy tracking that fit with the reporting mechanisms and practices of DPR and the mayor's office. As individual entities, the trees are human-scale at the time of planting (although they may grow as high as several stories over time). When trees are first installed on streets, they are 2.5–3 inches in diameter and approximately 10–15 feet high; reforestation trees are potted, only up to a few feet high, and can be handled by individual volunteers. Thus, trees invite direct, tangible, physical, and emotional interaction on an individual level (see also Jones 2014).

As part of their biology, trees interact with human and atmospheric domains in ways that some people have framed as *environmental externalities, benefits, and services*. Starting from this biological and material base, quantified data and modeling about the benefits and services of trees have become a core part of the discourse about the urban forest. First, perhaps most historically important across time and culture: tree limbs and leaves provide shade. Long before the techniques existed to estimate the impacts of shade trees on home energy bills or on mediating the effects of the urban heat island by lowering surface air temperature, people have sought out the cool air underneath trees. Second, trees are associated with cleaning water. In a rural context, forested lands are crucial to the protection of freshwater sources, as agriculture and residential development threaten the quality of surface and underground water. In aging cities dealing with combined sewer systems, the ability of trees (and the tree pits in which they are planted) to absorb and retain water helps keep additional water out of the combined sewage system and, therefore, lessens the likelihood that untreated sewage will be released into the surrounding waterways. Third, trees transform the visual landscape as they grow and are lauded as providing beauty. Numerous studies have demonstrated the impact of trees on creating walkable, pedestrian-filled streets and vibrant commercial areas, and raising property values. The aesthetic transformation of an area through tree planting has been documented with before-and-after renderings and images of MillionTreesNYC block plantings.

Sometimes the relationships among trees, humans, and the environment are not as straightforward. By taking in carbon dioxide and emitting oxygen, trees are associated with cleaning the air. However, this association is complicated by the fact that (1) as trees grow, they also emit other particles, such as volatile organic compounds, which are pollutants; (2) the shape of the urban tree canopy can affect the distribution of ground-level ozone and other pollutants; and (3) trees can also release pollen and other allergens that affect air quality and human health. The scientific, policy, and management arenas are actively exploring the relationship between a neighborhood's tree canopy and childhood asthma, for example. Although the science is far from settled, this relationship did not stop managers from instituting the TPH effort to concentrate tree planting in neighborhoods with high incidences of childhood asthma.

Despite their crucial role in sequestering carbon in rural forests, the role of urban trees in carbon sequestration is not foregrounded in the urban-planning discourse, for numerous reasons. The primary impact of urban trees on carbon emissions comes indirectly, via the reduction of energy used for home cooling (because of shade provisioning) and home heating (because of wind breaks) (McPherson et al. 2008). The direct sequestration of carbon in urban forests has a small impact because urban forests have a relatively smaller land area compared with that of rural forests; only large, mature trees sequester significant amounts of carbon, and many urban trees are quite small; and urban forests are often intensively managed, requiring the use of petroleum-based inputs (gasoline, fertilizer, pesticides) that change the carbon balance. Further, the way trees are handled after they die affects carbon storage—wood products can store carbon for quite some time, though currently few urban forest wood products are made—and landfilling, chipping, and mulching all have different effects on the way carbon is stored and released (McPherson et al. 2008; Ryan et al. 2010).

In order to survive, trees have needs for water, soil, sun, and space in particular configurations that define their ecological niche. Urban foresters have determined *rules-of-thumb and management practices* that can provide many of these basic needs to help ensure the longest possible survival of trees in the urban environment. These practices are followed throughout the actor-network of public managers, private contractors, and resident stewards, as described in chapter 3.

MillionTreesNYC resulted in major changes to tree procurement practices that will continue to influence plant quality and diversity in perpetuity. In addition, the first two years of life of a street tree are the most tenuous, so DPR requires private contractors to guarantee (and replace, if needed) their trees for two years. Ideally, this triggers the contractors to water these young trees on hot summer days—although when contractors are less diligent, they focus more on replacement than maintenance. After these two years, street trees have much higher survival rates but do benefit from ongoing care, including watering and pruning. While DPR provides its own maintenance, numerous stewardship programs have also been developed to enlist the help of resident volunteers in street tree care. The space and orientation of the tree pit/planting area also affects tree health and survival. While previously tree pits were three feet by three feet, DPR expanded that standard size and encourages pits to be as large as the sidewalk will permit, ideally five feet by ten feet, because of the observation that the larger the potential rooting volume, the healthier the urban tree. DPR has also experimented with locating trees off the curb, in the middle or the interior edge of the sidewalk, away from the hazard of car doors, although these planting arrangements are still nonroutine.

While CFH has routinized and refined many of its management practices in the PROW, NRG's reforestation and afforestation efforts continue to evolve through adaptive management. Because the work on invasive species removal and forest restoration began only in the 1980s, there has not been sufficient time to know conclusively which practices work best for the long-term health of the forest. As a result, the city is engaging with researchers to revisit old restoration sites to determine how they are now performing and to understand which management approaches yield a sustained impact over time (Johnson 2012; Simmons et al. 2016). NRG does have standard restoration practices to which it adheres, and these have been further codified and documented through the *Guidelines for Forest Restoration* (City of New York DPR 2015). Still, natural resource managers have demonstrated a willingness to work with academic researchers in testing different planting palettes and mixes of trees and shrubs in experimental plots across the city (see, e.g., McPhearson et al. 2010; Simmons 2010; Felson et al. 2013; Falxa-Raymond et al. 2014). Starting in 2012, a newly formed nonprofit, the Natural Areas Conservancy, began work on a citywide ecological assessment that took plot measurements of upland forest composition, health, and abundance in order to provide a comprehensive baseline for understanding the city's forests and to inform decision-making going forward.

The *material network of the urban forest* extends further backward in time and outward in space from New York City. Regardless of where they are planted, all of New York City's trees originate from commercial nurseries throughout the

Mid-Atlantic and Northeast region. These large-scale nurseries apply arboriculture expertise, labor, and material inputs including water and chemical pesticides. With the onset of tree procurement contracts, DPR began to purchase directly from nurseries rather than having private landscape contractors purchase trees as part of the planting process. In this way, DPR was able to control production of their own plant material, from cultivation methods to species palette, branching characteristics, size, and hardiness. Prior to arriving at the procurement nurseries on the East Coast, the trees that end up on New York City streets are purchased as "starts"—or tiny saplings—from large-scale tree farms, such as J. Frank Schmidt in Oregon. Trees are shipped across the country and the region on flatbed trucks, which requires all the inputs associated with long-haul shipping. The trees, shrubs, and herbs planted in forest sites all come from local stock that DPR either grows at the Greenbelt Native Plant Center or provides to area nurseries to grow.

Another key input, particularly for street tree planting, is fresh soil. While the excavated soil from a tree pit is put back in the ground, the process often requires the addition of fresh fill. This soil is acquired from greenfield—meaning uncontaminated, undeveloped—sites throughout the region. Research on how to amend and improve urban soils on-site without adding additional clean soil, such as through the use of phyto-technologies, is currently under way.

A fundamental material reality of trees that influences how they are planned for and managed and how they interact with the built environment is the fact that they *grow*—in many cases dramatically changing shape over time. On the PROW, the interaction between trees, sidewalks, people, and power lines is an intricate dance in a finite amount of space. Trained arborists from both DPR and utility companies prune trees to avoid overhead wires and select trees that work best for sites with wires. Citizen pruners also engage in street-level pruning (not around power lines) to maintain the health and appearance of trees and the accessibility of sidewalks. Particular tree species are selected, pit size is expanded, and sidewalks are repaired to address the raising of the sidewalk by growing tree roots. And the perception that tree roots will damage sidewalks, basements, and foundations is one of the areas of concern homeowners express in complaints to 311 (Rae et al. 2010). In addition to the spatial interactions on the PROW, this tree growth has a temporal dimension, as many trees take years or decades to reach full maturity. Planning for the duration of the campaign, DPR entered into contracted growing arrangements with nurseries in order to ensure a sufficient, high-quality stock of trees. In addition, investments into the urban built environment are being made for the benefit of future generations. Although young trees do make a visual impact upon planting, many of the benefits and services described above are not discernible until trees mature.

Finally, as living entities, trees *die*. Although they are being treated as infrastructure "like a stop sign" (respondent 27), a tree is materially different from a stop sign, and city managers have to address the mortality of their green infrastructure in the long term. Some critics have called for greater acknowledgment of the full lifecycle of trees in the PROW, encouraging DPR to leave standing dead trees on the street, rather than removing them and managing the streetscape to look like a "climax forest" (Hoffman Brandt 2012). However, standing dead trees in the PROW pose potential public safety concerns. Currently, when street trees die, they are cut down and chipped to use as mulch or landfilled (in the case of trees in the Asian Longhorned Beetle quarantine zones). Recycling and adaptive reuse of urban wood waste is currently being explored in a number of cities, such as Baltimore, although the practice is not yet widespread (Bratkovich 2001; Solid Waste Association of America 2002; Sherrill 2003). As of 2016, a range of partners including DPR, the New York State Department of Environmental Conservation, and the U.S. Forest Service's Forest Products Lab are in conversations about urban wood utilization and potential innovations in a "new urban forest economy."

The MillionTreesNYC campaign counted every tree that it planted in reforestation plantings toward the million tree goal, with the full understanding that some of these trees would be outcompeted in the forest setting. This tree death in the context of the forest is not a failure, as it is part of natural forest growth (as contrasted with the death of a tree in the PROW, which managers seek to stave off by every means necessary). DPR is conducting research on mortality in the reforestation sites, but mortality at those sites is not tracked as publicly as the sheer number of trees planted. Finally, though NYRP counts the number of trees it gives away to the public, it has no way of tracking how many of these trees survive. Overall, these examples complicate the meaning of the million tree goal—it was a crucial rhetorical device to build political and public support but does not reflect the full dynamics of a growing and changing urban forest.

Farms and Gardens as Assemblages

Urban farms are highly varied assemblages of the built environment, biotic actors, human labor, and institutions. I examine these assemblages, looking at both what claims are made about farms and gardens as well as what basic material conditions they require. In their sheer diversity of form and function, urban farms and gardens can serve as material touchstones for far-ranging claims. To some extent, then, urban agricultural sites and urban agricultural discourses co-constitute each other. There are, however, some basic inputs and resources that all urban farms and gardens require that shape the way in which agriculture is managed in the city and how it enters the policy arena. Sites vary widely in their qualities such as lot size, soil quality, water access, and sun/shade exposure, thereby producing a broad range of individual, organizational, and networked approaches to resource management. This assortment of forms and governance arrangements poses challenges to municipal policymakers who try to monitor, track, and understand the meaning and impact of these sites, which leads to the claim that there is insufficient data about urban agriculture.

Food provisioning is one of the dominant practices and discourses associated with agriculture throughout history and across space. Informed by agricultural knowledge, climate, and cultural traditions, people grow plants and raise livestock to provide food for themselves, their families, their communities, and market consumers. Efforts are currently under way to quantify, qualify, understand, and promote urban sites as valid and important sites of food production. More generally, these spaces are harnessed into broader discussions about food justice; food security; healthy eating; combating hunger, obesity and diabetes; and eating local. Through the prism of a food systems lens, which informs all of the current city-led food visions and plans, urban agricultural sites are evaluated for and understood as sites of local food production.

At the same time, much of the space within urban agricultural sites is *not cultivated for food production.* These spaces may include shade trees and nonedible plants but also pathways, seating areas, stages, picnic tables, *casitas*, and other structures. In particular, the community-gardening tradition in New York City is as deeply connected to social, cultural, educational, and recreational space as it is to agricultural space (Mees and Stone 2012). Numerous efforts attempt to catalog the multidimensional benefits of urban agricultural sites (Cohen et al. 2012). This push toward metrics and quantification is part of an effort to speak to policymakers and funders in order to garner their political and financial support. For example, one area of policy innovation is the Department of Environmental Protection (DEP)'s Green Infrastructure program, which offers grants for installations of green infrastructure technologies (green roofs, bioswales) that retain storm water on private land. Drawing upon data about the ability of urban farms to retain storm water, urban agriculturalists successfully advocated that DEP expand its categorization of green roofs to include rooftop agriculture as an eligible green infrastructure technology. Subsequently, the Brooklyn Grange rooftop farm at the Brooklyn Navy Yard was partially funded via DEP. On the other hand, leaders, publications, and research references repeatedly emphasize the sociocultural values of urban agricultural sites, particularly as sites of beauty, solace, inspiration, creative expression, cultural heritage, neighborhood cohesion, and memory (Baker 2004; Saldivar-Tanaka and Krasny 2004; Glover et al. 2005; Svendsen 2009; Stone 2009; Tidball et al. 2010; Mees and Stone 2012).

Gardens and farms are also prized as educational spaces where youth and adults can learn about the environment, ecology, the lifecycle of plants, collaborative planning, and management, particularly given the dynamism of the sites (Sheffield 1992; Sandler et al. 1995; Ferris et al. 2001; Fusco 2001; Marvy 2009).

Further, the *basic needs of plants for sun, space, water, soil, and labor* serve as challenges that land managers and institutions work to address—and that shape the practices of urban agriculture. Each of these needs/challenges is examined in turn. Across each of these domains, we can identify "fixes" that have emerged to meet the material needs of gardens and farms.

Sun and space. In the developed city, identifying wide, flat, sun-exposed sites is the primary challenge. It is what has driven the move to rooftop farming, as well as the experimentation with growing on temporary Housing Preservation and Development (HPD)–owned sites that are subject to future development. It has also driven the emergence of groups like 596 Acres, which provides online geospatial data about citywide vacant lots, places signage on vacant lots, and does community organizing to help support the adoption of lots. More formally embedded within government is the Urban Agriculture Taskforce, an effort to identify vacant city-owned land and assess its potential for use in urban agriculture.

Water. The need for water is a constant challenge for farms and gardens in the urban environment. It has led to the development of rainwater-harvesting systems, ranging from simple rain barrels to more complex cistern systems. Depending on capacity and interest, groups can implement these technologies on their own or can seek the assistance of nonprofits like GrowNYC, which has helped install more than eighty rainwater-harvesting systems citywide. DPR and DEP have developed an institutional arrangement to address water access for community gardens such that all groups registered with GreenThumb can receive permits and tools to gain access to nearby hydrants.

Soil. The need for healthy, contaminant-free, and fertile soil has led to a range of management practices, informal to formal. At the site level, gardeners engage in soil quality testing and add compost to their beds—enhancing the quality of soil over the course of seasons and years is part of the natural, accretive process of planting and tilling the land. In addition, GreenThumb and NYCHA provide assistance on how to build raised beds and offer soil and compost to gardeners. The Department of Sanitation (DSNY) is another crucial entity in this network because of its leaf litter collection and compost programs. Community gardeners strongly opposed cutbacks to these programs, which were an important subject at public meetings leading up to PlaNYC 2.0. Rooftop agriculture requires new technologies, such as lightweight soil mixes. With this new technology comes questions about the longevity of rooftop farming and the nutritional quality of plants grown in these new soil mixes.

Labor. Farms and gardens require human labor to thrive. Work is involved in planning sites, preparing soil, planting seeds and plants, harvesting crops, composting and managing waste, general maintenance, and holding events. This labor comes from a wide variety of participants and users, including volunteers and paid staff. In an urban context, these sites are managed by a broad set of civic, private, and public actors through polycentric governance arrangements that are less hierarchical and more bottom-up than those supporting the urban forest. One key moment when individual sites intersect larger nodes in the network is through the site visit, in which outreach coordinators and staff from nonprofits and municipal agencies visit gardens and farms to offer technical assistance, materials, information, programming, and support. These points of interaction in the field help reinforce ties between sites that are physically dispersed across the city. Nonetheless, the material forms, planting palettes, management techniques, and use of these sites vary widely in response to the localized expertise, preferences, and needs of stewards.

Seasons. Just as trees grow, mature, and die, so, too, do urban farms and gardens change over time, particularly in New York City in response to the seasons of the Mid-Atlantic United States. This seasonality informs the rhythm of how farms are managed. The main growing season is spring through fall (from roughly April to October) with high agricultural output throughout the summer and fall harvest season. As such, the large umbrella organizations that support urban agriculture and community gardening hold their seasonal events in the spring (seed and plant giveaways and sales), summer (tours, stewardship events), and fall (harvest celebrations). This seasonality has historically led to challenges in establishing school gardens in New York City, given that schools are not in session during the primary season of productivity. The school garden network Grow to Learn provides information and case studies about options for managing the garden with interns, staff, or summer school students or advises school gardeners to "let it be" by planting cover crops to restore the soil over the summer (Grow to Learn 2013). The winter season is a time when gardens and farms mostly go dormant, with only perennial plants, shrubs, and trees visible on the site and less human activity and use. Gardeners and farmers have implemented technologies to extend the growing season including cold frames, hoop houses, and more elaborate greenhouses. The nonprofit New York Sun Works has created educational and community greenhouses across the city, particularly emphasizing rooftop greenhouses. Private firms like Gotham Greens also use rooftop greenhouses to grow greens hydroponically for sale.

Understanding a farm or garden as an assemblage requires an in-depth analysis of how these various biotic and abiotic actors are enmeshed.[4] Moreover, generalizing about urban agricultural assemblages in New York City is challenging because of the amount and variability of these sites, with approximately 500 community gardens, 650 NYCHA gardens, 15–30 urban farms—depending upon the criteria used—as well as innumerable gardens in private yards (Bennaton 2009; Stone 2009; Ackerman 2011). Nonetheless, this section has highlighted some of the common patterns that arise from the material realities of these sites in order to begin to trace the role of nonhuman actors in urban agriculture policy and management networks. Land, water, and soil have voices in the policy arena, and they "speak" through the maps we make, the institutional arrangements we configure, the claims we make at public hearings, and the policies we choose to pursue or ignore. Paying attention to nonhuman actors helps us move from the social network diagrams of organizational ties presented in chapter 1 to an understanding of actor-networks as hybrid assemblages of humans and things.

Change from Above, Below, and In-Between

William Sites (2003, xix) notes, "The urban terrain is shaped from above, from below, and . . . from in between. Thus, at certain moments . . . international economic forces will be seen to contract the political space for national and local policy; at other moments, however, neighborhood-based actors emerge to influence the direction of city policies; at still others, a city crisis spurs and helps to shape major reordering of national urban priorities." Although Sites's focus was on housing and community development, we can apply his notions to the construction of urban nature. We can certainly identify key moments when the urban milieu shifts in response to an external shock, such as the 2008 global financial crisis. Yet these very shocks are also *always* being shaped from below—they are comprised of millions of practices by millions of actors that led to that crisis. As such, these forces occur in parallel, rather than in neatly segmented temporal stages or spatial scales. In both cases of urban forestry and agriculture, we see forces from above, below, and in-between shaping and reshaping the urban terrain. These shifts occur through the mechanism of sustainability plans but also through a wide range of practices that far exceed the scope of the plan.

From Above: Economic Changes

Changes in the global economy have clear effects on the creation and management of urban nature. Within the period of 2007–2011, New York's economy went from booming to experiencing a massive financial crisis and a subsequent global recession. In 2007, the City of New York was operating with a budget surplus and had been for the prior four fiscal years. As such, $199 million was included in the 2008 budget and $1.6 billion were committed across a ten-year capital budget for PlaNYC projects (ICLEI 2010b, 39). DPR, alone, received approximately $400 million for tree planting efforts. Similarly, nonprofits faced a period of relative prosperity, with corporations, philanthropies, and private donors making commitments to support urban environmental activities.[5] For example, the Mayor's Fund received a combined $10 million from Bloomberg Philanthropies and David Rockefeller in 2008; and NYRP secured Toyota, Home Depot, and BNP Paribas as lead corporate sponsors, starting in 2007, for the MillionTreesNYC campaign.

The 2008 global crisis triggered rising unemployment, declining consumption, and the failure of many of the prominent banks and financial firms that power large sectors of the New York City economy, with cascading effects on local environmental work. The municipal government faced declining city revenues starting in Fiscal Year 2008, and all of the city' agencies had to meet targets of 30 percent across-the-board budget cuts. These budget cuts helped trigger the shift within DPR from planting reforestation areas with employees and contractors to more heavily engaging volunteers as a cost-cutting measure, and also led to cuts to the maintenance budget for the care and pruning of street trees. It also led to a hiring freeze across DPR that limited the agency's ability to handle staff turnover, which became a more acute issue when the agency sought to restructure its NRG division after that group failed to expediently spend $11 million in PlaNYC capital funds. In addition to cuts to the municipal budget, federal cuts to the Department of Housing and Urban Development's (HUD) Community Development Block Grant program threatened the GreenThumb budget, which relies primarily on this source. Overall, interviewees felt that in lean times, funding for parks and open space is more likely to be cut than funding for police, fire, and education. In the political jockeying over the municipal budget, the environment can sometimes be seen as an amenity rather than a necessity of urban life (see also Brecher et al. 1993).

At the same time, this period saw a *growing* engagement of the philanthropic sector, high-level donors, and individuals in urban agriculture work. The Doris Duke Charitable Foundation began supporting urban agriculture research and analysis initiatives in the New York City region in 2010. The Community Food Funders Group is an affinity group of local foundations in the New York metro area that formed in 2011.[6] Chef Rachael Ray's Yum-o! Organization and chef Mario Batali's foundation both pledged funding support for the creation and development of the Grow to Learn school garden network in 2010–2011. At the individual level, many respondents speculated whether the challenges in the economy were (1) pushing young people more often to pursue nontraditional employment paths, including urban agriculture entrepreneurial ventures, or

(2) encouraging more people to engage in gardening and farming to supplement their food budgets. Regardless of whether that linkage *actually* exists, this discourse was prominent and widely circulating among activists, policymakers, and media alike; and it fit with Lawson's (2005) thesis that economic crisis and community gardening have been linked throughout American history.

The financial crisis also altered the process of sustainability planning. When PlaNYC was being updated in 2011, no capital funds were committed to its initiatives. This was a stark change from the prior plan, which municipal leaders and publications celebrated for not just being "a plan on a shelf" but for placing substantial city resources behind the initiatives. Although urban agriculture *was* included in PlaNYC 2.0, this inclusion was more of a symbolic gesture of good faith and intention rather than a commitment of tangible municipal support. Moreover, in 2007, city leaders explicitly stated that they sought for the initiatives of PlaNYC to be within New York City's jurisdiction, budget, and political capacity to implement—congestion pricing being a notable exception to this approach. In contrast, the 2011 plan included references to harnessing the energy and capacity of civil society in implementing the plan, touting the importance of the role of civic groups and individuals. This shift was a response to the critique of the process of PlaNYC as top-down but also a pragmatic approach in the face of declining municipal budgets.

Although PlaNYC 2.0 did not offer any additional financial resources, it is important to note that the DEP Green Infrastructure Plan was released in September 2010. This plan became an important source of funding both for bioswales in the PROW and traffic islands retrofitted for storm water retention, as well as private efforts including green roofs and rooftop agriculture through the green infrastructure grants program. It was the city's response to complying with Clean Water Act regulations from the Environmental Protection Agency and the state Department of Environmental Conservation, and it reflects the way in which municipal austerity did not trump ambitious planning processes that had been started years before and advanced incrementally via PlaNYC, the *Sustainable Stormwater Management Plan,* and the ongoing work of public bureaucrats.

From Below: Organizational Learning and Social Movement Change

This book has shown how policymaking is not a moment in time when goals are formulated, but rather an extended process of goal setting, implementation, learning, and revision. The urban forestry case includes both the iterative learning of professionalized organizations (public and private) as well as the development of the campaign in its evolving efforts to cultivate a broad base of public support and engagement. In contrast, the urban agriculture case is closer in its institutional form to a social movement, with some examples of professionalized learning embedded within that network. This movement also changed in terms of relationships and tactics among its members, with various coalitions forming and ending over time. I'll illustrate these claims with several examples.

First, DPR's reforestation efforts showed substantial evolution in their framing and execution. The stated goal in PlaNYC was to reforest two thousand acres citywide, an effort that DPR initially attempted to pursue via its $1 million design contract with the firm EDAW. Having combed the city's landscape through GIS and field-based research, DPR determined that this goal was not attainable and instead shifted its emphasis to the number of trees planted in reforestation efforts, which, in turn, helped solidify and cement the million trees figure as *the* tree goal of PlaNYC. Regarding implementation, reforestation was able to incorporate research plots into the design of the plantings. The plantings were also reorganized to have a substantial volunteer stewardship program, thereby enhancing the constituents interested in the care and maintenance of these forest sites.

Second, DPR showed an evolving understanding in how best to support tree survival via (1) altered planting practices and (2) various partnership strategies and outreach techniques to cultivate volunteer stewardship. Initially, critiques circulated about the provision of capital funds through PlaNYC without sufficient expense funds to cover the maintenance of this green infrastructure. As part of the professionalized conduct of urban forestry, DPR foresters implemented best practices in tree selection and installation to help support the longevity and health of trees on the PROW. DPR altered its terms with private contractors planting in the PROW, engaged in contract-growing arrangements with regional nurseries, and built in a high level of oversight and supervision. Beyond these large contractual shifts that were built into DPR's practices from the outset of implementing PlaNYC, there were many more fine-grained examples of constant tweaking during the campaign. DPR and NYRP both learned by doing, with feedback from foresters and field crews helping refine the practice of planting each successive season. In terms of stewardship, campaign leaders worked hard to draw in potential allies and critics through the advisory committee and to invite them to be a part of building a stronger, networked campaign that would leverage private and nonprofit resources. Starting in 2009, they built a Stewardship Corps among the largest nonprofit greening nodes in each borough, essentially subcontracting with these groups to deliver stewardship trainings and providing resources to support this effort. By 2011, MillionTreesNYC had developed both a finely honed volunteer experience for single-day tree planting events and opportunities for longer-term engagement of the public via the Natural Areas Volunteers and the TreeLC workshops and mini-grants.

Third, NYRP shifted its strategies away from an emphasis on direct-planting efforts and toward more tree giveaways to homeowners, working in partnership with community groups. Financial necessity partly drove this shift, as direct-planting efforts were substantially more costly than tree giveaways, and after securing the three lead corporate sponsors, NYRP was less successful in raising large-scale corporate funding for the continuation of the ten-year campaign. Another challenge was logistical and even political, as NYRP faced challenges in partnering with major public landowners like NYCHA in order to plant on their lands. This led NYRP to focus on planting partnerships with large-scale private landowners. In refocusing on tree giveaways, NYRP used the tree as a way to build rapport with residents and neighborhood groups—to "meet them where they are" by offering a free resource, in the hopes of engaging them in other sorts of environmental acts and on other landscape sites (including parks, gardens, and the waterfront). NYRP also sought to alter the perception that it was a "go-it-alone organization" (respondent 11). As such, the giveaway process was refined from reaching out directly to the public to partnering with community-based groups that would help get the word out to their constituents about the availability of and need to care for trees. Via the latter approach, NYRP was able to draw upon more of the nodes in the urban environmental network.

The social movement around community gardening and urban agriculture in New York City has evolved and shifted over the last forty years. With deep roots in shared (but diverse) experiences of disinvestment and decline, many community gardeners in New York City first became engaged in gardening in the wake of the 1970s fiscal crisis. Those involved in community gardening were the people who either could not or did not want to leave New York City during this time; there were many people of color, low-income residents, and people committed to their neighborhoods throughout all the many challenges they endured. Community gardening has rarely proliferated as a top-down, municipally led strategy; it thrives as a grassroots, participatory, and neighborhood-scale endeavor. At the same time, I note the important baseline material and organizational support provided to gardeners over the years by the GreenThumb program since its founding in 1978. And the nonprofit, umbrella organizations serving gardeners citywide have grown in their capacity and sophistication. While these municipal programs institutionalize and nonprofits professionalize, so, too, are new organizations and endeavors constantly forming and disbanding—demonstrating that the movement remains dynamic. Many of the newer organizations and coalitions (such as FSNNYC, Flip the Table, Food Action, and BFC) brought an increased attention to issues of food systems, beyond the original focus of gardens on neighborhood stabilization.

Other notable shifts have been the tactical changes from straightforward land stewardship, to more adversarial tactics, to negotiation and compromise. At the height of the "garden wars" in the late 1990s, when Mayor Giuliani bulldozed and attempted to auction off gardens, activists were in full-on oppositional mode to city policies. They worked to elevate the visibility of the problem via the media, staged protests, and engaged in street demonstrations, parades, and sit-ins at gardens. They called upon local, national, and international networks of allies, while individual residents formed bulldozer hotlines, and peer networks circulated information, rumors, and strategies. Then, nonprofits TPL and NYRP got involved to help purchase land for the preservation of several dozen gardens. Meanwhile, sympathetic public bureaucrats worked from inside the system to help gardeners however they could: as advocates, as information leaks to the media, and eventually as part of the settlement with the attorney general. Over the course of many years of pre- and post-settlement work, these bureaucrats across DPR, HPD, and the Department of City Administrative Services formed ties and mutual understanding; they worked across their institutional structures in support of their various missions, swapping lots for use as gardens or housing, and avoiding further crises on the scale of what had happened before. These bureaucrats learned to mediate between agency heads and city leaders—with certain sets of priorities—and residents, gardeners, and activists—with other, very different sets of priorities. The 2011 formation of the Urban Agriculture Taskforce out of the mayor's office, with all the city's landholding agencies and garden advocates serving in an advisory capacity, represented an important shift. One of the first meetings at City Hall drew humorous comments from participants about how previously they were protesting on the steps and how they now—quite literally—had a seat at the table.

However, these tactical shifts have not been across-the-board, as contestation, protest, and guerilla-gardening tactics remain in use. For example, when the settlement with the attorney general expired in 2010 and garden rules were being renegotiated, garden activists (particularly through the NYCCGC) began to mobilize again, to attend public hearings, to call upon the media, to pressure GreenThumb, DPR, and city leaders to offer strong protections to community gardens, and to engage in lawsuits. In essence, the relationship between garden advocates and public officials is a dynamic one, with advocates continuing to push for recognition and rights for gardens so that they are never again viewed as temporary or expendable spaces. At the same time, a new wave of activists *not* previously engaged in the long history of garden advocacy is bringing a different perspective to the abiding quandary of how to find space to farm and garden in the city. For example, the group 596 Acres presents a mix of street-level tactics, such as placing informational signage on vacant lots, community organizing, and digital mapmaking to engage residents in making claims on vacant public and private land. In some ways, the tactics of reappropriation of space have come full circle to some of those used in the 1970s, but in a very different historical and political economic context.

In-Between and All Around: Changes in Societal Values

Many respondents discussed their engagement in urban natural resources stewardship and sustainability policymaking within the context of longer-term shifts in societal values. This finding was surprising to me because it was not something that I sought out in this temporally tightly bound investigation. But it kept being repeated in respondents' understandings of why PlaNYC was happening *now*, or why urban agriculture was so popular *now*. This could not simply be chocked up to Bloomberg's power, nor to the networked savvy of nonprofit groups, nor to the social movement tactics of the grassroots—something at a more *societal* level was also afoot. Just as movements around feminism, civil rights, and marriage equality have progressed over time through normative shifts across the generations, so, too, have environmental awareness and pro-environmental values become more mainstream among younger generations. Caring about the environment is a broad issue, with many fissures and nuances within it. Nonetheless, we can detect that people who have grown up since the wave of environmental advocacy of the 1960s and policies of the 1970s have a different viewpoint from those who grew up prior to that period.

First, we can examine who is participating in municipal public service and what values they carry with them. In addition to the stability of leadership and expertise provided by long-time public agency bureaucrats, many respondents noted the importance of *young* staffers in City Hall, on the city council and in the executive agencies in bringing new ideas and perspectives about environmental issues. Bringing with them progressive environmental values, these young staffers are able to "push the envelope" on a wide range of policies and innovations around food systems, green infrastructure, and open space design, to name just a few issue areas that respondents mentioned. In addition, because sustainability requires cross-sector solutions and agencies working beyond the bounds of their conventional silos, individuals who are somewhat less entrenched in these existing bureaucratic structures may be freer to challenge or alter those structures.

Second, we see a rise in the number of young people engaging in self-driven, entrepreneurial, and community-based ventures around urban agriculture. Many of the most prominent examples of urban agriculture projects and leaders in the print media, online, in public visions and plans, and at public meetings and conferences come from a demographic of young, educated people opting into gardening and farming as professional and/or lifestyle choices, often associated with preferences around local, organic, seasonal, and artisanal food. These young people come to serve as "hipster archetypes" of what urban agriculture *is;* and this identification has led to a troublesome linking of urban agriculture and gentrification. This raises questions about what is lost when gardening shifts from being a neighborhood stabilization strategy in times of decline to a bourgeois, lifestyle choice. This archetype is, in turn, parodied and critiqued as necessarily incomplete and detrimental, in obscuring the decades of work of a very different demographic of people engaged in gardening and farming in the city, to much less fanfare. At the same time, it is important not to consider all newcomers to urban agriculture as an undifferentiated mass. First, there is considerable variation among the participants in this new wave of urban agriculture. And second, even when there is some surface uniformity by age, race, or class, these agriculturalists vary widely in their values. They might care about urban farming and gardening because of preferences about their individual food consumption, or because of broader worldviews and values of ecological sustainability, social justice, and economic self-sufficiency, as evidenced by the complex constellation of discourses associated with food, farming, and gardening. Only time will tell how these changes in societal values toward urban nature and sustainability continue to progress, evolve, diminish, or broaden.

This book unpacks the contemporary construction of several key components of urban nature in New York City. Attending to both discourse and nonhuman actors, we gain a richer understanding of the urban political process. At the same time, bringing a detailed analysis of urban politics together with a nature-society approach helps continue to advance our understanding of ecology as fundamentally political. Lessons learned and questions raised from this in-depth analysis of New York City's urban nature can be applied to (or juxtaposed with) other cities that are engaging in sustainability planning and investments in green infrastructure.

8

CITY AS ECOSYSTEM

Changing Form, Function, and Governance of Urban Socio-Nature

Nature is political *everywhere*, including—or perhaps particularly—in the city. Perspectives from human geography help us work to move beyond old binaries of nature and society as we examine the construction of urban nature as a set of discursive, political, and material processes. In the prior chapter, I synthesized key findings in each of these thematic areas across the two case studies. Still, it remains challenging to think with an *assemblage* mind—to avoid falling back on the categorical containers in which we put actors and things, politics and place, ideas and action. In this conclusion, I seek to correct that shortcoming by thinking holistically about these complex assemblages of urban socio-nature. Realizing the unique, context-dependent story I have shared, I also raise questions and observations from the New York City case for practitioners in other locales who are interested in supporting and catalyzing transitions toward sustainability. In so doing, I reexamine the way in which a "city of forests" and a "city of farms" has taken hold in the policy agenda, the stories we tell, and the transformations we make to our cities.

Sustainability planning is a key intervention into the governance of urban nature that brings with it institutional reorganization. It necessarily takes a long-term view that is not often adopted due to short-term political cycles. We see this long-term approach in the 2030 horizon of PlaNYC, as well as the commitment to update its goals every four years. Sustainability planning encourages interdisciplinary action by looking holistically at environmental systems from land to air to water to climate. The City of New York helped to break down administrative silos with the creation of the Office of Long-Term Planning and Sustainability as a research, coordination, and policy hub for sustainability issues. We can also find ample examples of interagency coordination in the pursuit of sustainability agendas, such as DPR and DEP's collaboration on the roll-out of green infrastructure efforts.

Sustainability—while certainly a flexible frame—is meant to balance economic, environmental, and social concerns. The aim of PlaNYC was to enable the "green growth" of New York City; however, it placed less emphasis on issues of social justice or social welfare. At a recent conference, Gary Machlis of Clemson University critiqued urban sustainability-planning efforts for this common bracketing out of social concerns: "Too often, the sustainability goals that are used ignore something: the alleviation of suffering. When a city measures its sustainability on the amount of bike paths it has and ignores its access to safe, warm housing in the winter for the homeless, or shelters for women in abuse, that's not a sustainable city. . . . I suggest a critique of all the sustainability metrics that are being used today" (Machlis 2015). Indeed, PlaNYC began with a focus on *land use* and municipal infrastructure; it was not a plan for reworking social services related to education, labor, immigration, homelessness, elder issues, and so on. The plan's most thoroughly social domain was the chapter focusing on housing—which proposed the development of both market-rate and affordable housing, with an emphasis on creating transit-centered hubs in outer boroughs. Where other social issues were considered, it was usually through a lens of improving public health or livability.

Yet if the livability of neighborhoods is disconnected from the people who actually reside in a place, it has the potential for environmental gentrification and displacement. This returns me to one of my interviewee's quips that PlaNYC "was a plan for a million more people working at Goldman Sachs." I do not share such a harsh critique for two reasons. First, although beyond the scope of my analysis in this book, I recognize that PlaNYC decision-makers attempted to integrate affordable housing into this sustainability vision. And second, I think that everyone can benefit from many of the plan's investments in common resources like air quality and open space. However, we will do better going forward to use a more thoroughly integrated approach to sustainability that includes social services in the scope of the plan from the outset, with a direct focus on issues of equity. Indeed, this is just what has occurred under subsequent administrations, with de Blasio's emphasis on affordability and equity in OneNYC, his reworking of PlaNYC. Going forward, disentangling and addressing issues of environmental gentrification are currently at the forefront of the thinking and work in research, policy, and activist circles.

Setting aside these—admittedly—large omissions related to how sustainability was defined, what did PlaNYC do and say about how urban land should be used and what form of urban nature should be advanced? The issues that made it onto the sustainability agenda or remained on the "cutting-room floor" provide a window into what was valued in these planning processes. In the case of New York City, we can see the way in which urban forestry was compatible—politically, spatially, materially, normatively—with the green growth aims of PlaNYC. Urban agriculture was less so—as it challenged basic assumptions about how space should function, be used, and be governed in the city.

City of Forests

Whether we consider it a product of interurban competition or policy transmission networks, there are fads and trends in urban environmental policymaking. Just as Los Angeles had a million trees program, so, too, did New York City, and later central Indiana and London as well. Ideas take hold, get vetted, become popular, and appear actionable. Models exist for replication, researchers conduct the science, journalists document the stories, and conferences and grey literature trade in best practices. Planting trees in large densities and wide spatial extents, in partnership between municipalities, nonprofits, and public volunteers, has come to appear in many locales as a commonsense sustainability strategy. How did that shift occur from advocates making arguments about trees to city officials pushing for their inclusion in plans? How did we come to embrace our urban forest?

Casting trees as green infrastructure that performs services makes them part of the engineered urban landscape. Using the models and language of ecosystem services, we can say that shade provisioning, water absorption, and heat reduction are all services of trees that we want in our high performance cities. And leaders like Bloomberg saw that planting trees—when combined with traditional grey infrastructure and built solutions—is a more cost effective and efficient way of managing ongoing environmental challenges like combined sewer overflow and urban heat island. We are seeing a shift in the mindset that takes the historic, engineered "sanitary city" and works to transform it into a "sustainable city" (Grove et al. 2015). In the case of water, this means reworking highly ingrained logics about how storm and waste water are retained, treated, and flow through the system. But in the case of trees, the shift required is not as large. It involves scaling up and tweaking already-existing practices: plant trees, more of them, in sites that will provide whatever co-benefits one is trying to maximize, and using best practices in their planting in terms of enlarged pit sizes, diversified species, and so forth.

But what often tips the ledger books into the black for urban trees in the eyes of policymakers are the ways in which they enhance quality of life and thereby raise residential and commercial real estate values. Captured in contingent valuation studies and well documented in the literature, the *feeling* of a tree-lined street makes a difference. For all of our precise modeling and valuation tools, it is the affective sense of awe and beauty that we feel under a cathedral of large canopy trees that lies behind much of what we are valuing. It is a person crossing the street to walk on the shady side. It is looking up through dappled sunlight. So we are compelled to plant trees—in the end—because we *like* cohabiting with them.

The street tree is really the darling of the urban forest in policy circles. It is the object that is modeled in STRATUM (later, iTree Streets); it is the primary focus of most municipal tree planting programs, volunteer shade tree commissions, and utility companies' pruning programs. Street trees visibly line our public thoroughfares and shade our homes. They are planted and managed as individual units with hundreds of hours of arboreal labor and care (see Bardekjian 2016). Municipal help lines, like 311, are filled with calls from the public either requesting street trees or complaining about them. Articulated as enhancing neighborhoods, street trees become fundamentally enrolled in the project of economic growth as pursued by city leaders, developers, and other landed institutions.

Riding on the coattails of these street trees are the less-visible trees in urban woodlands and "natural areas." While their benefits can similarly be modeled and quantified through forest plots and models like UFORE (later, iTree Eco), different arguments are leveraged on their behalf. These trees are largely in the domain of naturalists, park managers, birders, conservationists, and ecologists. While almost every urban dweller will encounter a street tree, trees in forested landscapes must be sought out. The practices and routines for managing these forests are still being developed through trial and error, adaptive management, experimentation, and organizational learning. It will take decades to understand whether the current practices of invasive removal and native replanting are fully successful in creating healthy, multistoried forests.

Even less visible still—or even sometimes maligned—are the "weed" trees like *Ailanthus* that grow as volunteers in vacant lots, backyards, and urban woodlands. These trees are explicitly not part of million tree planting campaigns (though they can be included in local tree ordinances and laws).[1] In the context of an urban woodland, trees like *Ailanthus* are invasive, which means managers aim to remove them and replant with other preferred species. Currently, the dominant management approach is one that prizes nativity and biodiversity, even in highly disturbed and fragmented forests that require massive inputs (human labor, herbicide) to maintain that nativity. I do not dispute the intrinsic value of biodiverse urban woodlands, but I wonder if there is space to open up a dialogue about when, where, and how nonnative weed trees provide benefits that we might celebrate—particularly outside the context of the woodland, when "weed" trees occur within the built matrix. For example, the tree across the street from me in a vacant lot is a weed tree. I gaze out on it from my writing desk. Its full canopy still inspires me; I mark the seasons by it; I admire the way it interrupts the roofline of the buildings. Similarly, the tree from *A Tree Grows in Brooklyn* was an *Ailanthus* tree; prior urban dwellers and writers have been inspired by this persistent tree. Indeed, DPR does plant nonnative trees on city streets, in recognition of their hardiness and ability to perform in that environment. Academic conferences and management seminars are filled with discussions about the merits and costs of native versus novel ecosystems—meaning our human-modified ecosystems of the Anthropocene, as this is very much an active area of current debate. It is important to remember that all trees provide both services and disservices; context is key in finding the "right tree for the right place."

The language of the "urban forest" encourages us to consider all of these street trees, park trees, urban woodlands, and backyard trees as part of a single system. And across that system, trees can be many things to many people. They are flexible objects through which a range of ends can be pursued. Overall, they have been cast as a key part of many cities' sustainability strategies. But sustainability for whom? Ecosystem benefits are not static things; they are co-constituted by relations between people and nature. Therefore, the benefits that trees can provide in an urban context depend largely on where they are sited, how they are planted, and how they are managed. Is beautifying streets the aim? Capturing storm water? Creating biological corridors for wildlife? Engaging youth in ecological learning? Providing "green jobs"? The answer, in many cases, is "all of the above" as savvy planners and bureaucrats seek to leverage the full suite of benefits. This language of maximization and co-benefits can, however, elide the hard choices that do occur, often behind closed doors. There is a great opportunity to use these potential tradeoffs and co-benefits as a starting point for democratic dialogue and shared decision-making. One such proposal is for the conduct of urban tree canopy prioritization efforts where stakeholders create their own weights on different desired outcomes, which can then be aggregated in a single spatial model to tell planners where to plant, and for which reasons (see Locke et al. 2013).

The emphasis on trees as infrastructure funded through infusions of capital dollars allows for bold leaps in the volume of trees planted. In other words, $400 million will plant a lot of trees—even in New York City. However, what this capital funding obscures is that trees are a living infrastructure that grow and die, requiring intensive ongoing management and stewardship. To truly create a sustainable forest requires capital and maintenance commitments; there is no shortcut around that. The choices about *how* that management occurs—through municipal employees, private contractors, nonprofit employees, public volunteers, or some mix thereof—have real consequences for what the urban forest *is* and how it performs. For the urban forest is not just the agglomeration of trees amidst the concrete and buildings—it is the ties that we create between ourselves and the trees. Fostering gainful employment, education, and engagement are all noble and worthwhile aims—but they require different strategies to pursue. These programmatic and maintenance strategies cannot be an afterthought in urban greening campaigns. Toward those ends, there is much that cities across the world can learn from community forestry's earlier aims of fostering empowerment, social justice, and civic engagement—before the turn to large-scale green infrastructure campaigns occurred.

So while it is exciting to see the way in which trees have risen to prominence in the policy dialogue, further evolution in thinking is required. Under what conditions—particularly climactic conditions such as in water-limited cities that historically were *never* forests—should that common sense be challenged?[2] While an expansion of the canopy made sense in New York City, there are other city contexts where the balance sheet is less clear. In those locales where it *does* make sense to plant more trees, I believe that we have an obligation to work to address environmental injustices in the distribution of our urban forests, where they exist.[3] Furthermore, we must move from tracking numbers of trees planted to a more holistic set of measures of the impacts made on social-ecological systems through tree planting and care. Just because the number of trees can be counted in a ledger doesn't mean it is the way that we should measure success. And yet, a tree planting goal provides a compelling storyline that can be packaged up and marketed—to policymakers, funders, and the public. What other storylines can we develop or goals can we set that make transparent—and exciting—why we are seeking to plant trees in the city? The MillionTreesNYC campaign staff understood the need not only to quantify trees planted, but also to qualify the meaning and significance of those trees, by sharing stories of community members and their trees online and honoring effective stewards through awards programs.

Perhaps we can take the best of both approaches—grassroots community forestry and citywide tree planting campaigns—in crafting a new way forward. The many actors involved in the governance of the urban forest provide different points of entry into the system. Municipalities can continue to fund and maintain their green infrastructure just as they support roads and bridges (though these, too, are often in dire need of investment!). But in order to embrace bottom-up engagement, they need to participate in new hybrid forms of collaboration and partnership with civil society. These partnerships provide opportunities for community input about which trees are planted, where, and when. When that happens, trees become a "widget" that can serve other community ends—coming together, making collective decisions, reclaiming space, or beautifying a neighborhood. The MillionTreesNYC campaign showed substantial evolution toward this way of thinking over the course of the campaign—most notably signaled by the creation of a Director of Stewardship position within DPR that develops programs across the range of site types on which stewardship occurs, from street trees, to parks, to wooded "natural areas." If community stewardship is built into the fabric of a campaign from the outset—not as an afterthought or as a labor force to come in and maintain the already-existing urban forest—it helps shape the form (and function) of that forest in new and surprising ways.

City of Farms

Cultivated gardens have always existed alongside our dwellings, but there is something fundamental about the historic divide between city and country that urban agriculture challenges. Traditionally, our large-scale, food-producing landscapes were located in rural hinterlands, and cities evolved as places for industry and exchange. Rethinking those spatial logics is radical in that it challenges some of our first principles about city planning and reworks our relationships between humans and nature. As we move into a fully globalized, postindustrial economy, however, the spatial form and pattern of cities continues to shift. In the ebb and flow of urban, suburban, and rural development and decay, there are always interstices of land that become "vacant" of previous uses and can be repurposed. We see productive food spaces occurring within urban and peri-urban areas—out of necessity, need, and innovation. Can we find continuities between a peasant garden along a railroad right-of-way in China, or Havana's network of farmers markets and organic farms in the face of scarce resources like capital and oil, or New York City's community garden network, or a hydroponic, green roof farm selling microgreens at Whole Foods?

The history of urban agriculture in America is particularly tied to the community-gardening tradition that has emerged over time in response not only to the basic impulse to grow food, but also in the context of various, systemic crises. The prior wave of farming and gardening in New York City aligned with the 1970s fiscal crisis and urban decline of the 1970s and 1980s. Community-led stewardship and control of these sites occurred in the wake of the retreat of capital and of the state. As such, these sites can provide food, but they also provide a space for engaging in other forms of community activities, for learning on the land, and for managing a *commons*. This tradition of the commons is crucial to understanding the legacy and current position of urban farms and gardens. Activists with ties to the community-gardening movement are not just advocating that land be given over to farming in the city, but also that it be kept in community control. While the site type may be similar (farm, garden), the governance structure is very different between a community-managed garden, a nonprofit farm, and an entrepreneurial farm. Community-based natural resource management occurs mostly outside the bounds of top-down managerial logics of city planning. Although it is variously enabled and constrained by city policies, actions, and agencies, gardens are not seen as the primary responsibility of any one public agency in the way that the urban forest is. Given this view, what forms of resources and support do community land managers need to best sustain their endeavors in the long term? How can we maintain community control without undernourishing or stifling it? These are key questions to the long-term sustainability of community gardening and other contemporary urban commons.

The latest wave of urban agricultural innovation and activism is driven less by localized expressions of urban crisis. Some attribute the rise in interest in local food and farming to changing preferences and lifeways. Others trace the ethic to perceptions of crisis at a global scale—from climate change, to food quality scares, to the obesity epidemic. The origins of this urban agriculture movement and how it intersects with broader food movements are certainly multiple and overdetermined, as discussed in chapter 5. In addition to long-term activists, gardeners, and farmers who have been working in New York City for decades, there is a new wave of people, some entrepreneurial in orientation, who are interested in farming in the city. They are farming on rooftops, hydroponically, and in temporary planters. They are claiming vacant lots, and they are taking over management of community gardens when prior generations of gardeners have retired, moved away, or passed on. In some cases, growers have less concern over the permanence of these sites than the prior wave of gardeners who fought for land tenure during the 1990s garden crisis; some growers are fine with a season or two of growing in temporary planters on a lot slated for development.[4] Other growers are less concerned about providing public open space or a neighborhood commons—and are managing private farms, including on rooftops. Finally, some are carrying on the efforts of prior generations in new ways, by both pushing for more land to be brought into production and fighting to preserve existing sites that are subject to development—as in the case of 596 Acres's and NYCCGC's activism or the Urban Agriculture Taskforce's advocating for a survey of vacant, available, city-owned land across multiple public agencies. This multiplicity of approaches shows the innovations and the shifting values around what it means to farm in the city.

Has urban agriculture risen to the level of a *sustainability fad?* In the case of New York City, not yet, at least not as compared to urban forestry. Urban agriculture has popular appeal amongst a certain set of the population, including community gardeners, farmers, and broader food activists, but it has not yet scaled up through a comprehensive citywide campaign like that which exists for tree planting. In other cities, however, we are beginning to see policy innovations, such as the urban agricultural zoning recently undertaken by San Francisco, that are sure to work their way through the policy-mimicry and competitive-city networks into other locales. The uptake varies, in part, with respect to the prevalence of vacant land and the real estate development pressures that exist in a city. In shrinking cities like Detroit, urban agriculture has progressed so much that it seems to be the benchmark city against which many urban agriculture practices are judged. In growing cities, there is a struggle for space over every single lot. However, if development pressures alone were the determining factor, we would not see the sort of policy innovation that we see in costly, gentrifying San Francisco Bay Area. Clearly, environmental and social values matter in setting our planning priorities as well.

What place does urban agriculture have in the broader discussion around local and regional food systems? The inherently multistaged and multiscaled nature of the food movement makes it distinct, albeit often in synergy with, the prior wave of community gardening. Notably, in thinking about *food* as the policy object, activists care about the ways food is processed, sold, consumed, and disposed, not just about how it is grown. This range of concerns opens up more possibilities for policy innovation in space-constrained cities. Food policies range broadly from funding for community incubator kitchens, to coupons that support low-income customers' purchases of healthy foods at farmers markets, to citywide composting initiatives, and much more. Whereas food was previously bracketed out of environmental regulations that focused on water, air, waste, toxics, and endangered species, those silos are beginning to break apart, and cross-cutting issues like climate change and food systems are shifting the way we think about the environment.

Food system activists care about local, urban food-producing spaces (farms, gardens), but they also care about forming ties to regional farmers who work outside of globalized corporate food systems. Thus, the policy responses they seek will need to operate similarly at multiple scales of city, region, state, and nation. The lack of coordinated regional planning authorities and the impact of federal agricultural policies on the economics of farming mean that the scope for maneuvering at the local scale is only partial. But local-scale solutions are always partial and incomplete. This is where we see the role of transnational activist networks that connect local, national, and global movements—with food justice and food sovereignty frames opening up spaces for collaboration between activists.[5] In New York City, we also see the role of planning and visioning documents to articulate possible future trajectories along many different scales and to move incrementally toward the implementation of policy changes. We see *FoodNYC* and *FoodWorks* not shying away from reaching beyond the municipal bounds. We see policy roadmaps that are the outgrowth of activists and public officials working together in dynamic and evolving networks. More so than urban agriculture in isolation, food system planning has the potential to go mainstream as a sustainability trend, bringing with it the broad range of interpretations and implementations that such a flexible approach contains.

City as Ecosystem: What Sort of Urban Nature?

In comparing the densely built, densely populated environment of the City of New York to, say, Yellowstone National Park, space is both *more* and *less* precious. It is more precious in that every square foot is accounted for, sought after, and contested—and real estate value is sky high. Every decision is costly; every corner has multiple potential uses. But it is also less precious in that it is open to change. As part of our heritage, national parks are managed largely as preservation landscapes. But in the case of cities, we accept a certain amount of dynamism and transformation in our built environment—buildings are erected and demolished, roads are planned and rerouted, entire coastlines are moved and rebuilt.

Forests and farms are landscapes that grow and shift as well. As a result, perhaps they can be compatible with our dynamic urban form? Can we find opportunities to weave trees, plants, water, soil, and air into our urban matrix in novel ways? As evidenced by the transformation in the public right of way through expanded tree pits and bioswales and the retrofitting of green buildings with vegetated walls and roofs, the current answer is a *partial yes*. Multiple actors from public agencies, to private developers, to individual residents are already pursuing these strategies. They range from low-tech "tactical urbanism" innovations like pop-up parks and parklets, to heavily financed and technically administered green building tax credits and certifications, to city ordinances mandating certain practices for trees and lawns.

Yet when most people think of "nature" in the city, they think of a particular form: the *park*. The park remains a crucial site of set-aside green space that serves a variety of uses and functions, including primarily recreation, relaxation, and socialization. Parks can be expansive in size and arranged in connected networks, such that they can also function as ecological corridors and biodiversity refuges. Many of our urban and national parks have common origins in a particular nineteenth-century understanding of conservation—they preserve a certain landscape aesthetic and promote certain uses over others.[6] I am in no way challenging the importance of the permanent preservation of green space in the city.

Just as we need preservation of historically significant buildings and districts, we benefit from ensuring urban green space in perpetuity.

What we can revisit are the ways in which our parks are governed, designed, managed, and used. Currently, we are seeing innovations in physical design and adaptive reuse of space—in notable projects like Van Valkenberg Associate's design for Brooklyn Bridge Park, built upon formerly industrial waterfront piers. We are also seeing governance arrangements that share financing, management, and decision-making between the public and private sectors—as in the case of New York City's Friends of the High Line that co-manages the site with the City of New York. Currently there are approximately forty public-private park conservancies in New York City alone. However, these public-private partnerships are not necessarily accessible across all neighborhoods and contexts. Like the arrangement behind MillionTreesNYC, they often thrive when the nonprofit partner has substantial resources, a visible public profile, and elite ties. Even though it entails a substantial role for civil society, it does not necessarily represent a democratization of the management of space. Thus, these public-private partnerships have been critiqued for setting up a "two-tiered" park system of haves and have nots (see, e.g., Harden 1999).

If we open up our parks even further to new governance forms and management priorities, we could see a broadening of the type of uses that are permitted and even encouraged in our parkland. We can already see the implications of community-based natural resource management in the community gardens that dot the city and how different they appear in form and function from other sorts of parklands. Can we imagine other instances where community decision-making is allowed and even encouraged on other site types? What if each neighborhood managed their parks differently according to local priorities, needs, and capacities? For example, one Brooklyn lot is a GreenThumb site that is managed by community residents with members and keyholders, but is currently programmed and managed as an "urban meadow" with wildflowers, a lawn, a performance stage—and no vegetable plots. So while the community garden is the quintessentially studied form of urban green commons, what other forms can we envision? At the same time, we can find examples of alternative means of financing greening projects, such as crowdfunding via online platforms like ioby.org. Peer-to-peer, resident-led fundraising on ioby has supported hundreds of projects from composting efforts, to school gardens, to bicycle-pedestrian transit pathways like Memphis's Hampline.

We can also rethink how and why we manage urban forests. What if we managed forests not only as urban conservation landscapes but as productive forests for foraging materials for food, medicine, and crafts? This is the notion behind Seattle's Beacon Food Forest, for example, which is managed for the berries, fruits, and other food products it yields. This concept also inspires artistic interventions, such as Swale, a floating food forest created by artist Mary Mattingly. What if urban forests were managed not only for their living trees but also for the forest products that can be produced after trees die—from lumber, to paper, to mulch, to biochar, to cellulose-based products, to biomass-to-energy? Numerous projects from small, urban lumber mills like Re-Co Brooklyn to large-scale biomass-to-energy efforts like District Energy of St. Paul's Combined Heat and Power Plant are beginning to alter the urban wood waste stream. Currently, much urban wood waste is chipped, mulched, or landfilled, but there are opportunities to develop alternative uses for this resource.

Ensuring that landscapes are managed sustainably and equitably in these novel forms of urban nature will not be easy. We have had centuries to develop the public administration structures, capital expenditure processes, and maintenance practices associated with urban parks. Similar attention needs to be placed on how we institutionalize new uses and interpretations of urban nature into our existing administrative structures, as well as on when structures need to be wholly revisited and reworked. It is so often that we see *physical design* of space leading the urbanist conversation, but there is as much opportunity (or perhaps more) to redesign *organizational and institutional structures* that shape and manage that space. For example, food pantries, volunteer harvesters, and homeowners with fruit-bearing trees are networked together via Toronto's "Not Far from the Tree," creating a whole new approach to managing urban fruit from trees on private land. Similarly, willing homeowners with unused backyards are connected with gardeners in search of space via Brooklyn's BK Farmyards. We can rework the relations of people, green stuff, and grey stuff through our organizational innovations like these.

While there will be bureaucratic hurdles and inertia along the way, at the outset we are limited only by our creativity of what sort of urban nature we can envision. For example, thinking about the urban forest as a multistage and multiscalar system, much like the local/regional food system, can guide us in reworking practices across the whole spectrum of sourcing, planting, management, and reuse. Thinking about what lessons we can take from community gardens to consider how other physical site types might be managed collectively or as a commons is another potential synergy. We can consider how green space, civic buildings, and housing can be redesigned in integrated fashions that help enable public engagement and community control of resources. Of course, these thought experiments lead to cascading questions that depend upon fundamental values, ideology, and goals. How will we decide when to manage as a commons, when to rely on public employees, and when to work via private contractors? By whom and for whom does the management of urban nature occur? Who "speaks" for the capabilities, needs, and preferences of the trees and the plants themselves? How can we create opportunities for meaningful encounters between human and nonhuman others in cities? Opening up these questions as points of discussion that are not fixed or settled can give some room to maneuver, some space for creative thinking.

And vice versa, these new and emerging interpretations of urban nature have much to learn about the existing processes of decision-making and management that are currently ensconced in the administration of urban green spaces (particularly parks, and to a lesser extent forests). By scrutinizing the current practices of sustainability policymaking and planning, we can identify potential pathways and inroads through which new ideas can find purchase. The central questions of this book explored how and why the urban sustainability agenda is set. These questions of *how* and *why* are codetermined and dialectic in nature. From these cases of urban forestry and agriculture, we observe whose voices were seen to matter, which issues were seen to matter, and how power operates in the construction of urban nature.

Ultimately, we must attend to how changes to urban nature affect the lives of residents. By changing physical land use, including by creating productive, socialized spaces such as community gardens, we are altering the relation between humans and nature. Changing the physical form of the city by planting trees and creating urban farms alters human-environment relations in subtle (and not-so-subtle) ways. This book makes a humble attempt to demonstrate some of the key relations between changes in policy and the socio-natural environment—now we must continue to observe how these changes affect people's lived experiences of place. Through the ongoing processes of observation, experimentation, learning, collective action, hard work, and reflection, perhaps we can move closer toward the inclusive, just, sustainable cities we seek.

Epilogue

FROM BLOOMBERG TO DE BLASIO AND BEYOND

On October 19, 2013, I participated in the largest single volunteer event that MillionTreesNYC organized. It was held around the one-year anniversary of Hurricane Sandy and located at Rockaway Community Park on the Rockaway Peninsula, which had been completely inundated during the storm. I was already quite familiar with Rockaway Community Park as I'd spent that summer helping to lead a large-scale study of park use and social meaning in the Jamaica Bay and Rockaway area. Our research team had documented the lack of use, the closed bathrooms, the cracked basketball courts, and the missing tennis nets at this particular park. We had spoken with a number of the residents in the NYCHA public housing development across the street who confirmed that the mosquitoes were so bad much of the summer that you couldn't even walk in this waterfront park. It was one of the most underused parks I had ever seen in New York City.

Given all of this, I couldn't help but think that twenty thousand new trees didn't stand at the top of my list for needs for this site. But there we were: planting trees, donuts in bellies, free MillionTreesNYC hats on heads. Yet, despite my skepticism, I somehow couldn't resist the magnetic, positive feeling that overtook me. Was it simply due to being outside? Was it doing something physical with my partner and friends? Was it seeing colleagues outside the office setting? Was it that feeling of handling baby trees with care, planting gestures of hope for future generations who might benefit from a biodiverse, healthy forest? Was it posing for pictures with the city's GreenNYC sustainability mascot, Birdie, and members of the Jamaica Bay Restoration Corps who were part of our summer research team? I've come to understand why so many of the volunteer tree planters became repeat attendees, as my colleagues had found in their research on the links between tree planting and civic engagement (Fisher et al. 2011, 2015). The tree became the widget to bring people together. What mattered was what surrounded those trees: *all of us.*

Beyond charitable assistance (or even the mutual aid of Occupy Sandy), the Rockaways also needed some opportunities for *fun* post-Sandy. The programming at this event was incredible; there was ample food, free giveaways, and a jubilant atmosphere. The event broke all records for attendance, and the trees were planted in a few hours. In addition to the planting, the organizers helped people engage with the Jamaica Bay waterfront. During our social assessment, one of the few parts of the park that we observed as being well used were the piers in the back where people would fish and crab. On the volunteer day, DPR's urban park rangers led nature walks and helped visitors look for waterfowl through loaned binoculars from the American Museum of Natural History. Later, after planting was completed, we had an enormous barbecue, with picnic tables set up and deck chairs laid about and people lingering and talking for hours. We spent more time socializing and eating than we did sticking trees in the ground. I was happy to see that participants were not just corporate groups or environmental "insiders" like me but also included numerous neighborhood folks who walked across the street when they saw something was finally happening in *their* park.

This October event was the last MillionTreesNYC volunteer planting of the Bloomberg administration. It showed all that had been learned over the course of the campaign about engaging the public, which was also demonstrated in the newly created position of director of stewardship at the DPR. Even though the campaign wasn't originally conceived with a community forestry ethos, the volunteer events, programming, and staffing had evolved to embrace that ethos. DPR now has an entire Stewardship Team that focuses full time on volunteer recruitment, engagement, and training—providing opportunities for the public to care for urban nature on streets, in forests, and in wetlands. Furthermore, through its location on the Rockaways, the planting also reflected the turn in attention toward post–Hurricane Sandy response and recovery.

If the planning focus of the "aughts" was *sustainability,* then it has become *resilience* in the "teens." Immediately following the October 2012 storm, Bloomberg announced the Special Initiative for Rebuilding and Resiliency (SIRR). The SIRR effort was an ambitious, interagency planning exercise that used many of the same OLTPS and agency heads who had been involved in emergency response and transitioned them to focus on planning for future resiliency. Indeed, several of the key leaders of PlaNYC—such as Seth Pinksy—were brought back in order to lead this aggressive planning effort. The successor report to PlaNYC was the *Stronger, More Resilient New York* report that SIRR released in June 2013 (City of New York 2013). The funding for the resilience projects came from a mix of federal, municipal, and private funds, some of which was already in hand and some of which remained proposed or projected (e.g., from the Federal Emergency Management Agency [FEMA] and HUD relief funds). Projects ranged across the spectrum of grey and green infrastructure. They included investments in transportation, communication, and emergency systems as well as projects focusing on coastal protection, changes to shorelines, and investments in green infrastructure. The resilience-planning agenda continues to move forward through high-profile funding competitions organized by the Rockefeller Foundation ("Rebuild by Design") and HUD.

Bloomberg made environmental planning and action a focal point of his legacy, whereas his successor, Bill de Blasio, has placed his attention on issues of social justice (particularly education and affordable housing). De Blasio received public criticism from environmentalists for taking a long time in appointing key agency heads related to the sustainability agenda, including the heads of DPR and OLTPS. OLTPS continued with an acting director through the summer of 2014, as part of a reorganization that resulted in the creation of a permanent Office of Recovery and Resiliency.[1] De Blasio appointed Mitchell Silver as the DPR Commissioner on March 21, 2014 (City of New York 2014b). Fittingly, Silver was a former head of the American Planning Association with a strong emphasis on equity planning.

The new mayor's agenda for parks and green space focused squarely on the issue of equity, instead of seeing green infrastructure as an engine for economic growth. Numerous media articles in 2013 and 2014 critiqued the two-tiered park system, with private nonprofit conservancies providing funding for Central Park and the High Line, whereas neighborhood parks in low-income areas struggled with maintenance and staffing issues. At this same time, New York State Senator Daniel Squadron proposed a legislative measure for revenue sharing across the park system, and his proposal caused discussion in media and policy circles. While de Blasio offered some initial support for such an approach, he ultimately has not yet pursued it (Rubenstein 2014). Instead, he allocated $80 million in capital funding targeted toward "neighborhood parks"—the less visible, less lavishly funded parks—in what came to be called the Community Parks Initiative. This capital expenditure still does not raise the baseline expense budget that covers the ongoing maintenance of parks, an issue that some members of the city council contest (Foderero 2014b). PlaNYC was institutionalized via the chartered creation of OLTPS and the requirement that all subsequent administrations continue updating the sustainability report every four years. In April 2015, de Blasio renamed this effort and released *One New York: The Plan for a Strong and Just City*—referred to as OneNYC—placing equity and affordability front and center in the sustainability plan.

Overall, how have urban forestry and agriculture fared in the post-Bloomberg era? In the case of urban forestry, the longer-term effects of PlaNYC have continued. The highly visible goal of planting a million trees continued on the momentum of its own rhetoric and determination. Notably, both Mayors Bloomberg and de Blasio, along with other elected officials, Midler, NYRP staff, DPR staff, local schoolchildren, and supporters celebrated the ceremonial planting of the millionth tree and the 1,017,634th tree. The event was a rare joint public appearance of the two mayors since de Blasio's election, and media outlets framed it as an explicit attempt to mend previously sour relationships between the two (Fermino 2015; Lemire 2015). Beyond the millionth tree, the capital commitments to DPR, changes to the procurement and contracting processes, and the development of new volunteer stewardship programming are all ways in which an enhanced commitment to the development of the urban forest has been institutionalized in New York City. DPR has continued to embrace volunteer engagement in the urban forest through its TreesCount!2015 tree census, which used a citizen science approach to inventorying all of the city's street trees, working with models developed by and adapted from the nonprofit TreeKit.

In the case of urban agriculture, existing programs have seen some renewed attention and support, and new investments are being made. In 2016, the USDA Farm Service Agency advertised for a new urban outreach coordinator to be based in Brooklyn—a sign of potential future directions for the agency. In addition to federal CDBG funding, GreenThumb's municipal budget was increased through city tax levy funding in fiscal year 2016, allowing for the addition of seven new outreach staff. De Blasio's Building Healthy Communities initiative focuses resources and programming on twelve underserved neighborhoods citywide and includes a plan to expand urban agriculture on NYCHA grounds in several of these neighborhoods. Community gardeners have also been a part of resilience-planning efforts, such as through the "Gardens Rising" project in the Lower East Side of Manhattan, which has received $500,000 in post-Sandy related federal funds via the New York State Governor's Office to undertake a feasibility study for maximizing stormwater capture in garden sites. At the same time, community gardeners and their allies, particularly through the NYCCGC, continue to fight for the preservation of their sites in the face of ongoing development. While nowhere near the scale of the 1990s garden crisis, more than a dozen gardens on HPD land have been slated for affordable housing development during the de Blasio administration. Regardless of mayoral position, engagement in urban agriculture is on an upswing, farm sites continue to proliferate, and networks continue to grow and strengthen. Whether that movement can achieve the critical mass necessary to scale up and affect citywide policy and natural resource management remains a question for the future.

Notes

INTRODUCTION

1. The phenomenon extends beyond urban centers, as the Nature Conservancy has a transnational "Plant a Billion Trees Campaign" that spans the United States, Brazil, and China: http://www.plantabillion.org/.

2. One example of a policy shift required city agencies to release a database of publicly owned lots that were potentially suitable for urban agriculture. As part of that process, by 2013, 135 municipal lots across 40 acres were identified as "potentially suitable" sites.

3. Ecosystem services are defined as the provisioning, regulating, supporting, and socio-cultural services that the natural environment provides that support human life. They are widely used in research and policy context and rose to particular prominence via the United Nation's *Millennium Ecosystem Assessment* that assessed the impacts of ecosystem change on human well-being. Monetizing the benefits of trees, parks, and farms follows in the tradition of ecosystem services research.

4. The issue of whether a rock, worm, or tree can be considered an "actor" in our accounts of phenomena remains an area of philosophical, theoretical, and empirical debate. The term "actant" refers to nonhuman agents, noting that they can play an active role in this world, though they may be different in kind than human beings.

5. As a primary research method, I conducted sixty-five semistructured interviews with leaders in sustainability planning, parks and recreation, forestry, and agriculture. I read and analyzed the plans, websites, and reports of PlaNYC, PlaNYC 2.0, and MillionTreesNYC to determine the discursive framing of urban forestry. Similarly, I reviewed *FoodNYC*, *Food in the Public Interest*, *FoodWorks*, and PlaNYC 2.0 to examine the discursive framing of urban agriculture. Finally, I conducted participant observation of organizational meetings and fieldwork involved in the implementation of urban forestry and urban agriculture practices over the course of 2007–2012, with follow-up fieldwork to document the end of the MillionTreesNYC campaign in 2015. For a complete methods appendix, see http://www.fs.fed.us/nrs/nyc/pubs/resources/City-of-Forests-City-of-Farms-Appendices.pdf.

6. I do not use the term *bureaucrat* pejoratively; rather, I use it to refer to unelected public officials involved in the technical development and implementation of policies and programs.

7. Global and transnational examples include U.N.-sponsored Agenda 21 and Local Agenda 21; the International Council for Local Environmental Initiatives (ICLEI), which is a nonprofit membership association for local governments; and the C40, which is a collaborative of forty large cities working to address climate change at the local level in partnership with the Clinton Global Initiative (UN 1992; ICLEI 2010a; C40 Cities 2013).

8. Local plans include New York City's PlaNYC2030, Chicago's Climate Action Plan, and Toronto's Environmental Plan, for example (City of New York 2007; City of Chicago 2008; City of Toronto 2000).

9. The effort to institute congestion pricing in Manhattan, modeled after a similar approach in London, was a notable counterexample. This effort required state approval but subsequently did not pass (ICLEI 2010b).

10. The original timeline of the campaign was 2007–2017. Because tree planting was running ahead of schedule and because of concerns about the mayoral transition, the timeline was revised in 2012 to end the campaign in 2015.

11. This two thousand–acre goal was subsequently revisited during the implementation of the campaign; see chapter three for more details.

12. The organization's initial focus was on improving the quality of parks in Northern Manhattan, including Fort Tryon Park, Fort Washington Park, and Highbridge Park. Later, during the community garden crisis under Mayor Giuliani, as described in chapter five, NYRP bought more than fifty gardens slated for auction as housing development sites, thereby expanding the organization's focus to include those gardens. With the MillionTreesNYC campaign, NYRP expanded to work on urban forestry projects citywide.

13. Stringer issued *Food in the Public Interest* in February 2009 and *FoodNYC: A Blueprint for a Sustainable Food System* in February 2010, while Quinn issued *FoodWorks: A Vision to Improve NYC's Food System* in November 2010. Both Stringer and Quinn were at one time mayoral candidates; Stringer went on to be elected to the office of Comptroller.

14. In thinking of urban nature as constructed, I draw upon a long lineage of philosophy and research pertaining to both nature-society theory as well as urban theory (see Smith 1984; Lefebvre 1991; Demeritt 2002 for key contributions on the production of urban space and the construction of nature from different philosophical vantage points). While constructivist approaches are highly varied, they all share an interest in questioning and examining knowledge claims; and some work in political ecology adopts a constructivist approach (see Robbins 2002 for a helpful review). At the same time, the return of the material in social theory can be seen as a response to "de-naturalized" approaches to human-environment relations that focused almost entirely on the discursive power of human thought, language, text, science, and action (see, e.g., Smith 1998; Castree 2005; Whatmore 2006; Bakker and Bridge 2006). In this book, I use the term "construction of urban nature" to refer to the political, discursive, and material ways in which urban nature is created as a concept and as a physical thing.

15. On the one hand, critical scholars have critiqued civil society on grounds of legitimacy, accountability, representation, and transparency (Syngedouw 2005; Heynen and Perkins 2007). Numerous authors have probed the accountability, representation, and self-determination of environmental and social civil society organizations, asking, "Who speaks for whom?" (see, e.g., Lake 1996; Getches and Pellow 2002; Bryner 2002; Schlosberg 2003, 2007). Gustavsson et al. (2009) question accountability in a case of a Swedish climate change governance network, arguing that "the blurring of hierarchies and indistinct roles and responsibilities within the network makes it difficult to exert accountability" (70). On the other hand, researchers have argued that civil society groups *do* have multiple sources of accountability, including members and boards of directors (Wapner 1995; Edwards and Hulme 2002). Some scholars claim that collaborative and/or networked governance can be good for democracy, cities, citizens, and the environment (Susskind et al. 1983; Susskind and Cruikshank 1987; Dryzek 1990; Healey 1997; B. Taylor 2009).

16. In sociology and political science, social-network research has looked at organizational alliances (Ansell 2003); ties among organizations that share members (Carroll and Ratner 1996; Cornwell and Harrison 2004); and the presence, structure, and effects of civic networks—which are "the web of collaborative ties and overlapping memberships between participatory organizations, formally independent of the state, acting on behalf of collective and public interests" (Baldassari and Diani 2007, 736). Recent scholarship has applied Social Network Analysis methods in examining: networks of environmental stakeholders (Prell et al. 2009); communication patterns and resource exchange (Crona

and Bodin 2006); links between social networks and resilience to climate change (Newman and Dale 2004); and organizational networks of urban civic environmental organizations (Ernstson et al. 2008), to name just a few examples.

17. The urban politics literature is not exclusive to the United States, as it has been explored and advanced by scholars in the United Kingdom, Europe, and elsewhere. I focus on American cities and the politics therein, noting the extensive cross-fertilization of ideas across the Atlantic. Building upon prior debates between pluralists and elitists, notions of urban regimes and the growth machine first originated in American scholarship; whereas explorations of governance and "the 3rd way" originated in the United Kingdom and Europe (see, e.g., Mossberger and Stoker 2001; Davies 2005; Jonas and Wilson 1999).

18. This topic was thoroughly taken up by regulation theory, which will be explored in the introduction to chapter one.

19. In examining the politics of a place, we can draw attention to dominant (or hegemonic) views of the environment and society, how they are concretized in institutions, by whom, and with what effects. Following the theorist and radical social critic Antonio Gramsci, hegemony is exercised when certain views come to be taken for granted as "common sense," even when these views might be detrimental to those that acquiesce to them (Crehan 2002; Lukes 2005; Loftus and Lumsden 2008). Ideologies are fixed (albeit temporarily) via social institutions, such as churches, schools, government agencies, and so on (Lees 2004; Loftus and Lumsden 2008).

20. I look beyond the formal plan, comparing and contrasting the different narrative frames associated with urban forestry and urban agriculture that are prevalent in local campaigns and implementation efforts. In the case of urban agriculture, which was not present in PlaNYC, I draw upon other sources to examine contemporary discursive framings. In particular, I analyze three policy reports: *Food in the Public Interest*, *FoodNYC: A Blueprint for a Sustainable Food System*, and *FoodWorks: A Vision to Improve NYC's Food System*. In the case of urban forestry, I examine references both directly in PlaNYC as well as in the MillionTreesNYC campaign website and materials.

21. Brand (2007) argues that "environmentalism has been a constitutive part of neoliberal urbanization" (618). Using Foucault's (1991) concept of governmentality, he argues that neoliberal discourses work to create "green subjects," claiming, "The environment is employed as a means of constructing citizens' sense of themselves and their obligations, in a manner perfectly attuned to the individualizing demands of neoliberal urban transformations" (Brand 2007, 628). While the issue of governmentality and green subject-making is certainly relevant and can be seen in the cases explored here, I primarily use a Gramscian understanding of power, hegemony, and ideology.

22. Latour offers three basic "litmus tests" for whether something is ANT. It must: 1. Treat nonhumans as actors; 2. Refrain from treating the "social" as an explanatory force; and 3. Aim to "reassemble the social" rather than simply deconstruct it (2005, 10–11).

23. ANT has been applied by geographers to the study of biodiversity conservation in the United Kingdom (Lorimer 2006); tubewell diffusion in India (Birkenholtz 2009); and practices of African elephant conservation and science (Whatmore and Thorne 2000). Robbins and Marks (2010) show that Latour's contributions are just one in a number of threads in the fast-growing field of assemblage geographies (dubbing Latour's approach "symmetrical," alongside Haraway's "intimate," Marx's "metabolic," and Mitchell's "genealogical" approaches).

24. To this already complex tale, I would add the need to consider abiotic actors, including buildings, sidewalks, and shadows.

25. *Assemblage* is a term that refers to the hybrid mix of actors, material things, and the relational network ties among them that comprise a phenomenon.

CHAPTER 1: GREENING NEW YORK CITY

1. Pincetl (2003)—drawing upon Cranz (1982), Foglesong (1986), and Rosenzweig (1983)—illustrates the lasting role of multisector partnerships in the history of urban park and open space planning and development. Writing about the contemporary Million Trees Los Angeles initiative, she calls the public-private collaboration a "coproduction" of the city and various nonprofits (Pincetl 2010). In a New York City case, Scobey (2002) traces the role of real estate interests and those of the landed elites in creating the institutions and landscapes that define the city (e.g., the street grid, Central Park), with long histories dating back to the early to mid-nineteenth century.

2. This section draws upon not only my key informant interviews but also data from the Stewardship Mapping and Assessment Project (STEW-MAP), which is a citywide census of civic environmental organizations that examines the social and spatial relations among urban actors involved in caring for the urban environment (U.S. Forest Service 2007; Fisher et al. 2012; Svendsen et al. 2016). I am one of the co-leads of this project. See the appendix for more details on methods: http://www.fs.fed.us/nrs/nyc/pubs/resources/City-of-Forests-City-of-Farms-Appendices.pdf.

3. Much SNA research is based on highly structured analysis of "complete networks," wherein all participants in a network are enumerated and surveyed, such that every tie between every single actor is documented, until a complete network matrix is collected. However, other recent research examines "ego networks"—the sets of ties closely linked to a set of egos (respondents)—and uses SNA qualitatively, as a way to begin to visualize a component of the network that one is studying. The limitation of the latter is that one cannot use many of the most powerful analytic tools to understand the structure and characteristics of the total network because one does not know the nature of ties from unsurveyed members of the network (Hanneman and Riddle 2005), yet qualitative SNA can be used to describe key actors and brokers in complex networks.

4. These include the Home Owners Loan Corporation of 1933 that led to mortgage redlining by race; the Interstate Highway Act of 1956, which subsidized the creation of a national highway system; the National Housing Act of 1934, which authorized the creation of the Federal Housing Administration to support the homebuilding industry; and the GI Bill of 1944, which similarly supported the homebuilding industry and provided veterans with access to homes via the Veterans Administration (Jackson 1985).

5. There have been many changes in the way in which race and ethnicity are tracked in the U.S. Census; in particular, questions related to Hispanic origin and multiple races have shifted dramatically over this period. Notably, the 2000 census was the first year in which respondents could select more than one race (Haub 2012).

6. "From 1961 to 1969 total spending increased and so did the share for redistributive purposes; from 1969–1975 total spending continued to rise, but there was less redistribution and more borrowing; from 1975 to 1983 there was reduced spending and reduced borrowing; from 1983–1989 spending again rose but it was allocated primarily to nonredistributive purposes" (Brecher et al. 1993, 9).

7. Although four of the five men confessed to the crime, they later recanted these confessions. Decades later, in 2012, DNA evidence exonerated the "Central Park Five" and led three of them to bring a lawsuit against the City of New York.

8. For the complete methodology of the study, see Fisher et al. (2012); see also the survey instrument at: http://www.fs.fed.us/nrs/nyc/pubs/resources/City-of-Forests-City-of-Farms-Appendices.pdf.

9. In terms of urban agriculture, 41.7 percent of groups in the study work on community gardens, while another 11.4 percent work on vacant lots, and just 6.5 percent work on urban farms. In terms of urban forestry, 23.8 percent of groups steward street trees, and another 17.3 percent work in the public right of way (which includes greenstreets and

plazas in the public right of way). Parks-based groups are prevalent (41.3 percent), as are groups that work on "natural areas" (17.3 percent), and groups working on school yards (11 percent). Groups also worked in and around buildings, including front and back yards (11.8 percent), apartment grounds (8.1 percent), courtyards (5.9 percent), and rooftops (4.5 percent), while 7.5 percent of groups referenced working specifically on green buildings (U.S. Forest Service 2007).

10. I did not ask interviewees to specify the type of collaborative tie—so these can include sharing resources, sharing staff, sharing information, working together on joint campaigns, and more.

CHAPTER 2: CREATING PLANYC

1. Flyvbjerg (1998), in his analysis of city planning in Aalborg, Denmark, called this sort of power-laden decision-making "rationalization" rather than rationality.

2. Chapters 4 and 6 continue the story of PlaNYC update and revision in 2011.

3. For example, the text of the PlaNYC initiative "Green the Cityscape" cites the precedent of the nonprofit Municipal Art Society's 1902 call to residents of Brooklyn Heights to green their neighborhoods by planting street trees and gardening in planters and window boxes (City of New York 2007, 38).

4. Interviewees speculated on the origins of this change in attitude. One cited Bloomberg's viewing of Al Gore's climate change movie, *An Inconvenient Truth,* in which the vulnerability of New York City to climate change was put in plain terms. Another noted that Doctoroff became a personal advocate for trees and greening after an environmental justice bike ride of the South Bronx. The origins and motivations are surely more complex than these single events, but it is possible that salient storytelling and firsthand experience played a part in mobilization.

5. 2008 N.Y.C. Local Law No. 17, N.Y.C. Charter §20.

6. Institutionalized political power rests heavily in the hands of the mayor in New York City, particularly since the Charter revision of 1989, which eliminated the former Board of Estimate that shared authorities with the mayor (Berg 2007). The mayor is the head executive of the city, and all of the municipal agencies report to that office, with commissioners of departments appointed by the mayor. He or she has the authority to create, reorganize, or terminate new executive agencies (City Charter 2009). The mayor also has broad authorities—with oversight from the comptroller and the city council—to set policies and to *propose* the capital and expense budgets for the city. The mayor's legislative counterpart is the city council which includes a public advocate and fifty-one other councilmembers elected to represent as many council districts (Berg 2007). The primary role of the council is to pass local laws and resolutions, as well as to serve as a check and balance against the power of the mayor. Thus, the council *reviews and approves* the budget proposed by the mayor, reviews the performance of city agencies, and has the power to call public hearings as it deems fit (City Charter 2009). The council also has authority over zoning, land use, and franchises—though the City Planning Commission also makes recommendations on zoning and land use (Berg 2007).

7. Also included under CFH are several programmatic groups: the Greenstreets program (renamed the green infrastructure division), which creates gardens in the public right of way; a trees and sidewalks group that repairs damage to sidewalks caused by trees; a GIS group that provides analytic capabilities; a horticulture group; a contract management group that oversees city contractors that install trees citywide; and a research division.

8. Other divisions within DPR besides CFH, NRG, and GreenThumb include Capital Projects, which is responsible for the physical development of new parks and the restoration of existing parks; planning; parklands; management and budget; revenue/

concessions; public programs; Green Apple Corps (a public service program); marketing and special events; the Urban Park Rangers (enforcement and education); and borough offices (NYC DPR 2013b).

9. The PlaNYC Advisory Board members were: *Elected officials:* Christine Quinn, speaker of the New York City Council; James F. Gennaro, City Council member and chair of the Committee on Environmental Protection; *Business/real estate community/design:* Steven Spinola, president, Real Estate Board of New York; Carlton Brown, COO and founder, Full Spectrum; Robert Fox, partner, Cook + Fox Architects; Elizabeth Girardi Schoen, senior director of environmental affairs, Pfizer, Inc.; Kathryn Wylde, president and CEO, Partnership for New York City; Daniel Tishman, chairman and CEO, Tishman Construction Corporation, chair Natural Resources Defense Council; *Environmental and community-advocacy representatives:* Marcia Bystryn, executive director, New York League of Conservation Voters; Peggy Sheppard, executive and co-founder, West Harlem Environmental Action Coalition (WE ACT); Andrew Darrell, regional director of NYC office, Environmental Defense; Ashok Gupta, program director of air and energy, Natural Resources Defense Council; Robert Yaro, president, Regional Plan Association; Elizabeth Yeampierre, executive director, UPROSE; *Academic community:* Ester Fuchs, professor, Columbia University's School of International and Public Affairs; *Philanthropic community:* Michael Northrop, program director, Rockefeller Brothers Fund; *Labor community:* Ed Ott, executive director, NYC Central Labor Council (ICLEI 2010b, 21).

10. Chapter 6 details the ways in which civil society and public actors worked outside the limits of PlaNYC to advocate for changes in local food and agriculture policy, which was then later reflected in a brief addition of food as a "crosscutting theme" in the April 2011 updated plan.

11. Many of the "brain trust" of environmental managers shared ties to the Yale School of Forestry and Environmental Studies and were colleagues who had been hired under the prior DPR commissioner, Henry Stern, and remained at the agency for several decades and through multiple changes of administration.

12. UTC requires remotely sensed or aerial imagery. At the time of the 2005 report, the accuracy of that imagery included a margin of error that meant that changes per year could not reliably be detected as distinct from error. Since that time, LiDAR imagery has improved the accuracy of detection, but it comes at significant expense to gather, process, and analyze (Ward and Johnson 2007; Secord and Zakhor 2007). Tracking progress by making projections based off growth and mortality rates was unsatisfactory to the leaders of PlaNYC.

13. The plan includes 127 stated initiatives, which are grouped into the goals that these initiatives help to achieve, and goals are further grouped thematically into chapters.

14. See also Schmelzkopf 1995, 2002; Smith and Kurtz 2003.

CHAPTER 3: CITY OF FORESTS

1. Although described as a ten-year campaign since its inception, in mid-2013 leaders updated the timeline to end in 2015 because planting was ahead of schedule. Leaders made every attempt to finish as much planting as possible while under the Bloomberg administration, with the campaign wrapping up over the course of de Blasio's first term.

2. Even with all the best sourcing, oversight, and installation guidelines in place, trees thrive when they have a level of maintenance that goes beyond what the municipality and its contractors could provide. As such, multiple stewardship programs were developed to engage the public in tree care—see chapter 4.

3. One respondent said this shift was due to changes in accounting and responsibility for trees planted on other municipal agencies' lands—which had previously been a "grey

area" in the campaign. DPR had been responsible for planting on parkland and NYRP for planting on private land—but how should plantings on other city agencies' land occur? Eventually, it was clarified that DPR would be responsible for coordinating all city agency plantings. Another said it was due to a decrease in NYRP's responsibility for planting on private lands, which DPR could absorb in increased reforestation plantings. In part, this was due to NYRP's realization that their direct planting practices were much more costly to execute than they had anticipated. Whatever the rationale, it led to decreased planting responsibility for NYRP in the end. A DPR spreadsheet documenting the final planting tallies from October 21, 2015, showed that DPR planted 745,600 trees, other city agencies planted 20,493 trees, and NYRP and other partners planted 233,907 trees.

4. When asked about what kept DPR accountable to the goals of PlaNYC (and the million trees goal in particular), interviewees identified employees' personal obligation, savvy leadership, hierarchical chain of command, frequent numeric reporting requirements—but most importantly, *the funding itself.*

5. Capital funds can be used for multiyear physical investments that will last over five years, whereas expense dollars go to cover the labor and maintenance to care for those investments. According to a 2013 report by New York City's Independent Budget Office, "Technically speaking, a 'capital project' involves the construction, reconstruction, acquisition, or installation of a physical public improvement with a value of $35,000 or more and a 'useful life' of at least five years" (NYC IBO 2013, 2).

6. The number of gardens purchased by NYRP varies in different reports, depending on how adjacent lots are counted. The number listed on NYRP's website is fifty-two gardens (http://www.nyrp.org/About—accessed on October 19, 2012).

7. Called "the piggybank" by one interviewee, the Mayor's Fund is intended to seek private funding for city initiatives (respondent 13). While it was previously used infrequently for one-off cultural events, such as parades, it was significantly expanded in scope under Bloomberg.

8. For a discussion of how trees as actants play a role in these practices, see chapter 7.

9. 2010 N.Y.C. Local Law No. 3, N.Y.C. Admin Code §18–107; see https://www.nycgovparks.org/rules/section-5 for DPR rules.

10. In the case of fruiting and flowering trees, the city plants only male trees in order to prevent dropping berries and fruits on city sidewalks—a position that is sometimes questioned by groups that are interested in promoting foraging and local food forests.

11. DPR also collected requests via phone, email, website, and letter from individuals, civic and business groups, community boards, and city council members.

12. TPH Neighborhoods are Hunts Point and Morissania in the Bronx, East New York in Brooklyn, East Harlem in Manhattan, the Rockaways in Queens, and Stapleton in Staten Island.

13. With 311 tree request printouts and GIS-based street maps showing tree requests, foresters drive and walk the city streets and assess planting locations. They look for conflicting infrastructural elements (such as driveways and crosswalks, overhead wires, adjacency to stop signs, hollow sidewalks, or underground vaults). Barring any of these major problems that prohibit planting a tree, foresters are instructed to locate trees wherever feasible. Foresters mark planting sites with spray paint and call utility companies to mark utility locations to avoid invisible infrastructure conflicts. Planting sites are then bundled together geographically and put out to bid for planting contracts.

14. Foresters have latitude in how large to make a pit, though the minimum recommended width is 32 inches, and the recommended size is 5 feet by 10 feet—with a minimum of 39 inches of clear sidewalk that must be maintained after planting (NYC DPR

2010, 19, 21). A street tree contractor noted that in his experience the average pit size was formerly 19 square feet and is now 42 square feet.

15. The only situations where an entity other than DPR and its contractors legally plants a street tree are: (1) occasionally a neighborhood association or business improvement district (BID) can raise the money to plant their own trees, or (2) in the case of new development or substantial renovation. PlaNYC institutionalized a change to the zoning rules that requires developers to purchase and install one new street tree for every twenty-five feet of street frontage (City of New York 2007). Though the BIDs or developers do these plantings, DPR is still responsible for issuing the permits and thus has an oversight role in monitoring new trees. Finally, in a limited number of cases, residents informally plant trees in tree pits "guerilla style," though this practice is formally discouraged by the city, and often foresters remove and replace these trees if they encounter them.

16. According to WNYC: "But while the trees' mortality rate is not unusually high compared to other cities, the Parks Department encourages New Yorkers to do their part to take care of them. According to the department, the two-year mortality rate was 6.7 percent for street trees planted in 2009. That rate—3.4 percent annualized—puts New York favorably in the company of Denver (2.4 percent, according to an internal audit); West Oakland (3.7 percent, according to a published study); and Los Angeles (4.4 percent, according to a yet-to-be-published study by the U.S. Forest Service). WNYC obtained more recent data, of trees planted in the spring of 2011, and determined a higher mortality rate: 12.1 percent over two years, or 6.2 percent annualized. (That cohort was the only complete data set obtained through a Freedom of Information Request.)" (Ye and Schuerman 2014)

17. Small trees were used in these reforestation sites to facilitate both site preparation and volunteer engagement. These small trees were planted four feet off center and were expected to grow quickly for rapid forest succession.

18. NRG developed operational routines for procurement, restoration, and planting in natural areas. Similar to the contracted growing arrangements developed for street trees, reforestation trees are purchased directly from a nursery. The main difference is that these trees are much younger and are purchased as "whips" in containerized pots, ranging in size from twelve inches up to five feet in height—with a desired minimum size of twenty-four inches. Native species from regional (200–400 miles) seed sources are preferred and used in the majority of cases. Prior to tree planting, invasive plants are removed from the site (through cutting, uprooting, mowing, and herbicide), and storm damage is addressed. After a site is cleared, crews work quickly to close the canopy by planting "whips" in close proximity to one another.

19. This partnership model was developed after the first few seasons of unsuccessfully attempting to give away free trees to people at events organized directly by NYRP. Once the leadership reoriented these events to be in partnership with local community-based organizations, with tree giveaways as a carrot, the demand for trees far exceeded the supply.

CHAPTER 4: BEYOND PLANTING

1. Within municipal agencies of the City of New York, the term "green infrastructure" is often used to refer to landscape interventions that help improve water quality by retaining water and mitigating combined sewer overflow. Following in the tradition of the EPA's definition of the term, green infrastructure is an alternative to the traditional "grey infrastructure" used in storm water management; and the term is central to the 2010 DEP Green Infrastructure Plan. The original PlaNYC document did not cast much of the tree planting as "green infrastructure" per se, though the PlaNYC 2.0 update did show DPR and DEP efforts as linked. In environmental management circles more broadly, the term

can refer to any natural infrastructural technology or solution that also provides ecosystem benefits (including climate adaptation, mitigating urban heat island, food provisioning, and more). I use the term here in this broader sense.

2. The growing number of nonprofit environmental organizations is not unique to New York City and has been documented as a national phenomenon (Straughan and Pollack 2008). With that growth comes an added capacity for innovation and public programs, but also the potential for competition for scarce resources and public attention (for one introduction to the relationship between resources, competition, and strategy, see Saxon-Harrold 1990).

3. In the early stages of the campaign, Majora Carter, a famous advocate for environmental justice, MacArthur "genius award" winner, and ex-director of the Sustainable South Bronx organization, publicly critiqued the MillionTreesNYC campaign. Carter had a history of sometimes adversarial relations with Mayor Bloomberg, concerns about the process behind PlaNYC, concerns about MillionTreesNYC's commitment to tree maintenance and stewardship, and an interest in cultivating green jobs. The advisory committee sought to "put the resentment aside and get things going" (respondent 20). To that end, Sustainable South Bronx was listed as a participant in the advisory committee, and MillionTreesNYC partnered with them in their green jobs training program.

4. Often, community gardeners were actively recruited to help serve as volunteer stewards of street trees, and some community gardens were identified as locations that might benefit from adjacent or surrounding tree planting. Some garden advocates took issue with the assumption that gardeners' labor could or should be recruited and noted that not in all cases did gardeners *want* trees planted nearby. GreenThumb developed outreach and training methods that acknowledge gardeners' role in managing their spaces and did not assume tree planting from the outset.

5. The organizational landscape surrounding natural areas continued to evolve as well. In 2013, a new nonprofit, the Natural Areas Conservancy, was created. This group emulated the concept of the Central Park Conservancy and the Prospect Park Alliance, serving as a private "friends of" group to support natural areas citywide.

6. Now called the green infrastructure division, the Greenstreets program transforms traffic triangles in the streetscape into green, planted areas—often with an emphasis on storm water capture.

7. Having participated in MillionTreesNYC volunteer plantings and thoroughly enjoying it, I recognize the dissonance between the abstract notion of my labors being harnessed and the felt, affective experience of planting trees as a voluntary, leisure, or civic engagement practice. Other researchers have studied the motivations and experiences of volunteer stewards engaging with the campaign (Fisher et al. 2011, 2015).

8. In 2011, Darin Johnson, who had been one of the key strategic leads in conceiving the entire MillionTreesNYC campaign, left NYRP. A series of leadership changes occurred at the programmatic level over the course of the campaign, with multiple changes in the MillionTreesNYC director position, development position, and forestry, horticultural, and field staff. NYRP's most significant transition occurred with the departure of Executive Director Drew Becher and the installation of his successor, Amy Freitag, in 2010. Becher developed a reputation for bold moves in reshaping environmental organizations through his track record in Chicago, Washington, D.C., and New York, as he was successful in growing budgets, hiring staff, and starting high profile public campaigns. Yet he was also known for a heavy-handed leadership and communication style. He was highly successful in upward accountability to funders and board members, but less so with downward accountability to members, employees, and partners. One of the ways this manifested was his greater focus on the number of trees planted than on the educational or stewardship programs that were put in place to support the maintenance of those trees.

In contrast, Freitag was initially viewed as more of an "insider" to DPR, as she previously was deputy commissioner for capital projects at that agency. Some of those involved hoped that this insider view would mend some of the tensions or miscommunications between the two sides of the campaign. She had a professional background in landscape and design and was invested in the sustainability of sites via the new planting practices of NYRP. Freitag's senior vice president and close collaborator, Deborah Marton, joined NYRP in 2011—and later went on to succeed Freitag as executive director. Marton had a similar track record and history of working with DPR on the design and management of New York City's environment.

CHAPTER 5: GROWING IN THE CITY

1. In addition to its mandate to develop affordable housing, HPD was subject to the directive issued by the Office of Management and Budget that it raise revenue by disposing of properties on its rolls through development or auction.

2. The number of gardens purchased by NYRP varies in different reports, depending on how adjacent lots are counted. The number listed on NYRP's website is fifty-two gardens (http://www.nyrp.org/About—accessed on 19 October 2012).

3. In total, eighty-six gardens on DPR land were listed as "*Parks Open Space*"; seven gardens on other public agency land (non-HPD) were listed as "*subject to development/ no development planned*"; and one hundred gardens on Department of Education (DOE) land were listed as "*DOE Open Space.*" All of these sites were designated to be maintained as gardens, offered the opportunity to participate in the GreenThumb garden registration process, and were required to go through environmental review and the Uniform Land Use Review Procedure (ULURP) if they were offered for disposition or development. Another 198 gardens were listed as "*offer for preservation*" either via GreenThumb or the nonprofit land trusts discussed above. Another 110 gardens were listed as "*subject to development,*" with the memorandum setting forth a protocol for public notification and review and seeking alternate sites for displaced garden groups. Finally, twenty-eight gardens that had already completed land-use and environmental review were listed as "*immediate development*" sites that could be sold by the city (Attorney General 2002).

4. After the timeline of this study, in fiscal year 2016 under the de Blasio mayoral administration, a significant expansion to the GreenThumb staff was made, with seven new staff and a budget increase of $421,000.

5. In fact, when asked to identify public, nonprofit, and business partners with which they work, several of my interviewees mistakenly identified GreenThumb as a civic group, revealing how differently it is perceived from other city agencies.

6. http://nyccgc.org/2010/08/recap-nyccgc-rally-parks-dept-public-hearing-regarding-proposed-new-rules/.

7. Some advocates push for community-run markets as opposed to Greenmarkets, which they felt were located in affluent areas and given top-down support by the city. A report by Manhattan borough president Scott Stringer laid out several critiques of the regulations and fees surrounding the operation of community-based farmers markets (Stringer 2011). Others did not feel the critiques against Greenmarket were valid, noting in particular the way in which Greenmarket provides affordable produce compared to supermarket competitors and participates in programs to encourage low-income consumers to shop at the market.

8. For example, Target funded the renovation and maintenance of community gardens in New York City, working with high-end landscape architects, facilitated by the nonprofit NYRP. While some welcome the support of corporate donors, others caution that if design and renovation are managed by a third party, then a community garden no longer functions as a community-managed open space but becomes more of a sponsored

"pocket park," without the associated benefits of community control and neighborhood cohesion. Similarly, Planters Peanuts sponsored the development of urban gardens as marketing opportunities; they sponsored a garden on NYCHA public housing grounds in the Lower East Side of Manhattan.

9. See http://www.nyc.gov/html/om/pdf/2012/otf_report.pdf.

10. This estimate was based on Barron et al.'s (2010) analysis of 2002 data from the Freight Analysis Framework. According to the authors, "In order to construct a macro-view of the New York City food system, we used the Freight Analysis Framework (FAF), a national dataset managed by the U.S. Federal Highway Administration that estimates commodity flows and related freight transportation activity among states, regions, and major international gateways. While the FAF dataset cannot be used to identify where food products originate or are consumed, it is a useful tool for analyzing the volume, dollar value, and transportation mode of food commodities flowing into, within, and out of the New York City region" (Barron et al. 2010, 10).

11. In March 2011, however, Heifer stopped their programming in this region to focus on other sites around the globe and the United States.

CHAPTER 6: CITY OF FARMS

1. Quinn and Stringer were both at one time running for mayor. Stringer dropped out of the race in November 2012 and ran for city comptroller, to which he was elected. Quinn was often described as a "frontrunner" but lost in the primary election in September 2013 and retreated for some time from politics.

2. It remained, however, a position of prominence until the last charter revision in 1989 that eliminated the Board of Estimate, considered removing the borough president altogether, and ultimately ended up reducing the powers of that office significantly (Berg 2007).

3. While it is clear that the food system is highly interconnected, I have chosen to focus only on the recommendations related to production, as these are most closely tied to supporting urban agriculture.

4. DCAS is essentially the "landlord to the city," responsible for owning and operating city facilities and leasing city-owned land to private enterprises. It is also the renamed version of the previously existing Department of General Services, under which the original Operation GreenThumb was created in 1978. However, not since that time has this agency been heavily engaged in enabling or facilitating urban agriculture or gardening, so this local law will involve city agencies in new ways.

5. 2011 N.Y.C. Local Law No. 48, N.Y.C. Admin Code § 4-208; 2011 N.Y.C. Local Law No. 49, N.Y.C. Admin Code § 27-306(c); 2011 N.Y.C. Local Law No. 50, N.Y.C. Admin Code § 6-130; 2011 N.Y.C. Local Law No. 51, N.Y.C. Admin Code § 6-317. To download the full text of the local laws, visit http://legistar.council.nyc.gov/Legislation.aspx.

6. There was an equally—if not more—prominent public critique of the plan's failure to fully address solid waste issues. Solid waste was subsequently incorporated as a chapter in PlaNYC 2.0. Many food advocates made note of this absence from PlaNYC and identified the connection of waste policies to food systems in the post-consumption phase.

7. Recognizing the complexity and interconnectedness of the food system, this table focuses only on initiatives focused on strengthening *food production*. Thus, it does not cover issues like distribution (including farmers markets) and post-consumption (including composting) unless explicit reference is made linking these policies to production.

CHAPTER 7: CONSTRUCTING THE "GREENER, GREATER" CITY

1. Of course, these narratives did not emerge wholesale out of the ether in these planning documents. They use storylines, tropes, evidence, and argumentation that are created and propagated via diverse sites, including television, print, and online media; peer-to-peer networks; social movement ties; celebrity engagement with issues; published research and fiction; and many communication venues.

2. Community forestry focuses on trees as a means rather than an ends and is epitomized by programs like the Urban Resources Initiative (URI) of New Haven, Connecticut. In this model, local residents and neighborhood groups use greening in order to achieve goals of neighborhood stabilization, beautification, recreation, education, and so on. Residents have a high degree of control over what trees are planted where and when and are provided with both material and technical assistance by URI staff and interns (Murphy-Dunning 2009). Similarly, programs like the Greening of Detroit shift power by hiring local residents and paying them for watering and caring for trees. Indeed, one respondent criticized the fact that MillionTreesNYC focused so much attention on volunteer stewardship and its expensive MillionTreesNYC Training Program, rather than having a broad-based, entry-level hiring program for tree care.

3. One can think of community gardens and urban agriculture as overlapping circles in a Venn diagram: not all gardens are agricultural or food-producing sites, and not all urban agriculture sites are managed as community gardens; however, there is some overlap between people and sites across these two categories.

4. The deeply situated, ethnographic approach that is best used to understand the micropolitics of site management (see, e.g., Glover 2003; Baker 2004) is more detailed and site-specific than the citywide scale of my research questions.

5. Nationally, across all sectors, philanthropic giving increased from 2006 to 2007: "Individuals and institutions made $306.69 billion in charitable donations and pledges in 2007, a 1 percent increase on an inflation-adjusted basis over the $294.91 billion given in 2006" (Strom 2008).

6. It includes Doris Duke, NorthStar Fund, Jessie Smith Noyes, Rockefeller Foundation, Merck Family Fund, New World Foundation, and Surdna Foundation.

CHAPTER 8: CITY AS ECOSYSTEM

1. New York City's current tree restitution law takes into consideration the ecosystem services and benefits that all trees under DPR jurisdiction provide—including nonnative trees. This tree restitution law does not apply to private property. Tree replacement value is determined on the basis of size, condition, species, and location (2010 N.Y.C. Local Law No. 3, N.Y.C. Admin Code § 18-107; see https://www.nycgovparks.org/rules/section-5 for DPR rules).

2. I thank Shawn Landry and Tenley Conway for organizing the "Trees and the City" panel at the American Association of Geographers meeting in 2016 where we discussed these critical issues. My copanelists Stephanie Pincetl and Christopher Boone pointed out the important differences in city conditions across different geographic locales and across long temporal scales (see also Avolio et al. 2015).

3. Current research by Kristen Schwartz et al. (2015) found a correlation between higher urban tree canopy and higher income in seven U.S. cities, leading the authors to claim "trees grow on money."

4. Yet I see repeated examples of projects that began as "temporary" interventions but came to be perceived or valued as "permanent" by the participants due to the ongoing deployment of labor and deepening levels of place attachment.

5. For example, the Brooklyn Food Conference in 2012 demonstrated dialogue among people working for a minimum wage in food service sectors in New York, farmers organizing for labor and immigration rights in Florida, and activists working to save native seed in India.

6. Erika Svendsen points out the linkages in the thought, discourse, and practice between Theodore Roosevelt and Frederick Law Olmsted in the creation of national parks and urban parks, respectively (Svendsen 2013). Nate Gabriel demonstrates the way in which nineteenth-century urban parks were crucial sites and spaces that served to articulate the divide between the capitalist city and the noneconomic, "natural" space of the park, thereby creating "urban subjects" whose appropriate use of parkland is for leisure (Gabriel 2011).

EPILOGUE

1. SIRR was incubated out of the OLTPS office under Bloomberg but later reorganized under the newly created Office of Recovery and Resiliency by de Blasio.

References

Ackerman, Kubi. 2011. "The Potential for Urban Agriculture in New York City." Columbia University Urban Design Lab Report. Accessed Nov. 28, 2012. http://www.urbandesignlab.columbia.edu/sitefiles/file/urban_agriculture_nyc.pdf.

Amin, Ash, and Nigel Thrift. 2002. *Cities: Reimagining the Urban.* Cambridge: Polity Press.

Anderson, L. M., and H. K. Cordell. 1988. "Influence of Trees on Residential Property Values in Athens, Georgia (USA): A Survey Based on Actual Sales Prices." *Landscape and Urban Planning* 15: 153–64. doi:10.1016/0169-2046(88)90023-0.

Angotti, Tom. 2010. "PlaNYC at Three: Time to Include the Neighborhoods." *Gotham Gazette,* April 2010. Accessed May 7, 2010. www.gothamgazette.com/article/Land%20Use/20100412/12/3239.

Ansell, Christopher K. 2003. "Community Embeddedness and Collaborative Governance in the San Francisco Bay Area Environmental Movement." In *Social Movements and Networks,* edited by Mario Diani and Doug McAdam, 123–44. Oxford: Oxford University Press.

Attorney General of the State of New York. 2002. "Memorandum of Agreement." State of New York by Eliot Spitzer, Attorney General of the State of New York ("State"), and the City of New York by Michael A. Cardozo, Corporation Counsel of the City of New York ("City"), Sept. 17, 2002. New York, NY. Accessed Nov. 28, 2012: http://www.nyc.gov/html/om/pdf/community_gardens_agreement.pdf.

Avolio, M. L., D. E. Pataki, T. W. Gillespie, G. D. Jenerette, H. R. McCarthy, S. Pincetl, and L. Weller Clarke. 2015. "Tree Diversity in Southern California's Urban Forest: The Interacting Roles of Social and Environmental Variables." *Frontiers in Ecology and Evolution* 3 (July): 1–15. doi:10.3389/fevo.2015.00073.

Bachrach, P., and M. S. Baratz. 1962. "Two Faces of Power." *American Political Science Review* 56: 947–52. doi.org/10.2307/1952796.

Baker, L. E. 2004. "Tending Cultural Landscapes and Food Citizenship in Toronto's Community Gardens." *Geographical Review* 94 (3): 305–25. doi:10.1111/j.1931-0846.2004.tb00175.x.

Bakker, K., and G. Bridge. 2006. "Material Worlds? Resource Geographies and the 'Matter of Nature.'" *Progress in Human Geography* 30 (1): 5–27. doi:10.1191/0309132506ph588oa.

Baldassari, D., and M. Diani. 2007. "The Integrative Power of Civic Networks." *American Journal of Sociology* 13 (3): 735–80. doi:10.1086/521839.

Barbaro, Michael. 2008. "Bloomberg Encourages Staff to Watch the Clock." *New York Times,* Aug. 17, 2008. Accessed July 31, 2013: http://www.nytimes.com/2008/08/18/nyregion/18clocks.html.

Bardekjian, Adrina. 2016. "Towards Social Arboriculture: Arborists' Perspectives on Urban Forest Labour in Southern Ontario, Canada." *Urban Forestry and Urban Greening* 19 (1): 255–62. doi:dx.doi.org/10.1016/j.ufug.2015.10.014.

Barrett, W. 2007. "All Wet—Bloomberg's Man Dan Doctoroff Has an Answer for Rising Seas: More Coastal Condos!" *The Village Voice,* March 13, 2007. Accessed Jan 2, 2013: http://www.villagevoice.com/2007-03-13/news/all-wet/.

Barron, Matt, Brian Goldblatt, Claire Ho, Rebecca Hudson, Dana Kaplan, Erika Keberle, Cullen Naumoff, Caren Perlmutter, Zachary Suttile, Cameron Thorsteinson, Debora Tsien, Lani Wild, and Meghan Wilson. 2010. "Understanding New York City's Food Supply." Program in Environmental Science and Policy at Columbia University's School of International and Public Affairs. Report prepared for the New York City Mayor's Office of Long-Term Planning and Sustainability. 92 pp. Accessed Dec. 7, 2012: http://mpaenvironment.ei.columbia.edu/news/documents/UnderstandingNYCsFoodSupply_May2010.pdf.

Bealey, W. J., A. G. McDonald, E. Nemitz, R. Donovan, U. Dragosits, T. R. Duffy, and D. Fowler. 2007. "Estimating the Reduction of Urban PM10 Concentrations by Trees within an Environmental Information System for Planners." *Journal of Environmental Management* 85 (1): 44–58. doi:10.1016/j.jenvman.2006.07.007.

Beckett, K. P., P. H. Freer-Smith, and G. Taylor. 1998. "Urban Woodlands: Their Role in Reducing the Effects of Particulate Pollution." *Environmental Pollution* 99: 340–60. doi:10.1016/S0269-7491(98)00016-5.

Bellush, Jewel, and Dick Netzer. 1990. "New York Confronts Urban Theory." In *Urban Politics New York Style,* edited by Jewel Bellush and Dick Netzer, 3–26. Armonk, NY: M. E. Sharpe, Inc.

Benepe, Adrian. 2010. "Protecting Gardens and Communities." *New York Post,* Aug. 11, 2010. Accessed Nov. 21, 2012: http://www.nypost.com/p/news/opinion/opedcolumnists/protecting_gardens_and_communities_GbyjZBlRqX5ivBMzFehC2L?listcomments=true#comments.

Bennaton, R. 2009. "The Re-Greening of Public Housing: New York City Housing Authority Garden and Greening Program, New York, NY." In *Restorative Commons: Creating Health and Well-being through Urban Landscapes,* General Technical Report, edited by L. Campbell and A. Wiesen, 233–47. Newtown Square, PA: U.S. Department of Agriculture, Forest Service, Northern Research Station.

Berg, Bruce. 2007. *New York City Politics: Governing Gotham.* New Brunswick, NJ: Rutgers University Press.

Bettencourt, Luis M.A., and Geoffrey B. West. 2011. "Bigger Cities Do More with Less." *Scientific American* (Sept. 2011): 52–53. doi:10.1038/scientificamerican0911-52.

Birkenholtz, T. 2009. "Irrigated Landscapes, Produced Scarcity, and Adaptive Social Institutions in Rajasthan, India." *Annals of the Association of American Geographers* 99 (1): 118–37. doi:10.1080/00045600802459093.

Black Urban Growers. 2012. "In Our Story: Stories of Black Food Sovereignty." Panel of speakers: Lorrie Cleavenger, Ray Figueroa, Melanie Samuels, Michelene Browen, and Nicole Taylor, at GreenThumb GrowTogether (March 31, 2012).

Blaikie, Piers, and Harold Brookfield. 1987. *Land Degradation and Society.* London: Methuen.

Borgatti, S. P., M. G. Everett, and L. C. Freeman. 2002. "Ucinet for Windows: Software for Social Network Analysis." Harvard, MA: Analytic Technologies.

Born, Branden, and Mark Purcell. 2006. "Avoiding the Local Trap: Scale and Food Systems in Planning Research." *Journal of Planning Education and Research* 26: 195–207. doi: 10.1177/0739456X06291389.

Bradshaw, Matt, and Elain Stratford. 2005. "Qualitative Research Design and Rigour." In *Qualitative Methods in Human Geography* 2nd edition, edited by Iain Hay, 67–76. Oxford: Oxford University Press.

Bragdon, David. 2011. "New York City as a Sustainable City." Columbia University Earth Institute Panel (April 7, 2011).

Brand, P. 2007. "Green Subjection: The Politics of Neoliberal Urban Environmental Management." *International Journal of Urban and Regional Research* 31 (3): 616–32. doi:10.1111/j.1468-2427.2007.00748.x.

Brash, Julian. 2011. *Bloomberg's New York: Class and Governance in the Luxury City.* Athens: University of Georgia Press.

Bratkovich, Stephen. 2001. *Utilizing Urban Trees: Ideas from across the Country.* U.S. Department of Agriculture, Forest Service Northeastern Area. NA-TP-06-01 (Oct. 2001). Accessed Aug 19, 2013: http://www.na.fs.fed.us/spfo/pubs/misc/umt/.

Braun, Bruce. 2005. "Environmental Issues: Writing a More-than-Human Urban Geography." *Progress in Human Geography* 29: 635–50. doi:10.1191/0309132505ph574pr.

Brdanovic, Davorin. 2009. "Gardens for Peace and Reconciliation" In *Restorative Commons: Creating Health and Well-being through Urban Landscapes,* General Technical Report, edited by Lindsay Campbell and Anne Wiesen, 248–65. Newtown Square, PA: U.S. Department of Agriculture, Forest Service, Northern Research Station.

Brecher, Charles, Raymond D. Horton, Robert A. Cropf, and Dean Michael Mead. 1993. *Power Failure: New York City Politics and Policy since 1960.* New York: Oxford University Press.

Brenner, N., and N. Theodore. 2002. "Cities and Geographies of 'Actually Existing Neoliberalism.'" In *Spaces of Neoliberalism: Urban Restructuring in North America and Western Europe,* edited by N. Brenner and N. Theodore, 2–32. Oxford: Blackwell.

Brownlow, A. 2006. "An Archaeology of Fear and Environmental Change in Philadelphia." *Geoforum* 37 (2): 227–45. doi:10.1016/j.geoforum.2005.02.009.

Bryner, Gary. 2002. "Assessing Claims of Environmental Justice." In *Justice and Natural Resources: Concepts, Strategies, and Applications*, edited by Kathryn Mutz, Gary C. Bryner, and Douglas S. Kenney, chapter 2. Washington, D.C.: Island Press.

Budiansky, Stephen. 2010. "Math Lessons for Locavores." *New York Times,* Aug. 19, 2010. Accessed Nov. 28, 2012: http://www.nytimes.com/2010/08/20/opinion/20budiansky.html?_r=0.

Buettner, Russ, and Ray Rivera. 2009. "A Stalled Vision: Big Development as City's Future." *New York Times,* Oct. 29, 2009.

Bulkeley, H. 2005. "Reconfiguring Environmental Governance: Towards a Politics of Scales and Networks." *Political Geography* 24 (8): 875–902. doi:10.1016/j.polgeo.2005.07.002.

Bumiller, Elisabeth. 1995. "Guarding the Turf, Stepping on Toes; Henry Stern, Passionate and Blunt, Champions the City Parks." *New York Times,* July 23, 1995. Accessed June 12, 2013: http://www.nytimes.com/1995/07/23/nyregion/guarding-turf-stepping-toes-henry-stern-passionate-blunt-champions-city-parks.html?pagewanted=all&src=pm.

Burrows, Edwin G., and Mike Wallace. 1999. *Gotham: A History of New York to 1898.* Oxford: Oxford University Press.

Burt, R. 2007. "Secondhand Brokerage: Evidence on the Importance of Local Structure for Managers, Bankers, and Analysts." *Academy of Management Journal* 50 (1): 119–48. doi:10.5465/AMJ.2007.24162082.

Campbell, Lindsay, and Morgan Monaco. 2014. "MillionTreesNYC: The Integration of Research and Practice." MillionTreesNYC and U.S. Forest Service. http://www.nrs.fs.fed.us/pubs/45580.

Campbell, Lindsay K., Erika S. Svendsen, and Lara A. Roman. 2016. "Knowledge Co-production at the Research-Practice Interface: Embedded Case Studies from Urban Forestry." *Environmental Management* 57: 1262–80. doi:org/10.1007/s00267-016-0680-8.

Canals, L.M., S.J. Cowell, S. Sim, and L. Basson. 2007. "Comparing Domestic versus Imported Apples: A Focus on Energy Use." *Environmental Science and Pollution Research* 14 (5): 338–44. doi:10.1065/espr2007.04.412.

Cardwell, Diane. 2010. "Six Stories above Queens, a Fine Spot for a Little Farming." *The New York Times,* May 13, 2010. Accessed Oct. 31, 2010: www.nytimes.com/2010/05/14/nyregion/14farm.html?ref=urban_agriculture.

Carroll, William K., and R. S. Ratner. 1996. "Master Framing and Cross-Movement Networking in Contemporary Social Movements." *Sociological Quarterly* 37: 601–25. doi:10.1111/j.1533-8525.1996.tb01755.x.

Caro, Robert. 1975. *The Power Broker*. New York: Random House.

Castree, N. 2005. *Nature*. New York: Routledge.

Centers for Disease Control and Prevention. 2009. "Obesity among Low-Income Preschool Children." Fact Sheet based on the 2009 *Pediatric National Surveillance System*. Accessed Dec. 7, 2012: http://www.cdc.gov/obesity/downloads/PedNSSFactSheet.pdf.

C40 Cities. 2013. "About C40: Sustainable Action on Climate Change." Accessed Aug. 2, 2013: http://www.c40cities.org/about.

City of Chicago. 2008. *Chicago Climate Action Plan*. Accessed May 9, 2010: http://www.chicagoclimateaction.org/filebin/pdf/finalreport/CCAPREPORTFINAL.pdf.

City of New York. 2014a. *PlaNYC Progress Report 2014*. Accessed 28 Aug 2014: http://www.nyc.gov/html/planyc/downloads/pdf/140422_PlaNYCP-Report_FINAL_Web.pdf.

——. 2014b. "De Blasio Administration Appoints Mitchell Silver as Parks Commissioner" Press Release, Mar. 21, 2014. Accessed Aug. 28, 2014: http://www1.nyc.gov/office-of-the-mayor/news/100-14/de-blasio-administration-appoints-mitchell-silver-parks-commissioner/#/0.

——. 2012a. "Mayor Bloomberg Releases New Wetlands Strategy to Protect and Improve City Waterways" Press Release 186–12, May 20, 2012. Accessed June 11, 2013, and no longer available: http://www.nyc.gov.

——. 2012b. "Patricia E. Harris: First Deputy Mayor." Accessed Oct. 19, 2012, and no longer available: http://www.nyc.gov.

——. 2011a. *PlaNYC Update 2011: A Greater, Greener New York*. Accessed Nov. 7, 2012: http://nytelecom.vo.llnwd.net/o15/agencies/planyc2030/pdf/planyc_2011_planyc_full_report.pdf.

——. 2011b. "Mayor's Management Report: Preliminary Fiscal 2011." Feb. 2011. Accessed Aug. 31, 2011: http://www.nyc.gov/html/ops/downloads/pdf/mmr/0211_mmr.pdf.

——. 2009. "New York City Charter." As amended Dec. 18, 2009. Accessed Aug. 9, 2013: http://www.nyc.gov/html/charter/downloads/pdf/citycharter2009.pdf.

——. 2007. *PlaNYC: A Greater, Greener New York.* Accessed May 7, 2010: http://www.nyc.gov/html/planyc2030/downloads/pdf/report_open_space.pdf.

City of New York Department of City Planning. 2011. "NYC2010: Results from the 2010 Census" (Mar. 2011). Accessed June 10, 2013: http://www.nyc.gov/html/dcp/pdf/census/census2010/pgrhc.pdf.

City of New York Department of Education. 2012. "About Us." Accessed Nov. 13, 2012: http://schools.nyc.gov/AboutUs/default.htm.

City of New York Department of Health and Mental Hygiene. 2012. "Working with Farmers Markets." (Oct. 23, 2012). Accessed Dec. 5, 2012: http://www.nyc.gov/html/doh/html/cdp/cdp_pan_health_bucks.shtml.

City of New York Department of Parks and Recreation. 2014. *Guidelines for Urban Forest Restoration.* Accessed Feb. 25, 2016: http://www.nycgovparks.org/greening/natural-resources-group/publication.

——. 2013a. "A Timeline of New York City Department of Parks & Recreation History." Accessed June 13, 2013: http://www.nycgovparks.org/about/history/timeline.

——. 2013b. "People" Accessed June 13, 2013: http://www.nycgovparks.org/about/people.

——. 2012. "Natural Resources Group." Accessed Nov. 7, 2012: http://www.nycgovparks.org/greening/natural-resources-group.

——. 2010. "Forester Handbook." Central Forestry and Horticulture Division.

City of New York Department of Sanitation. 2012 "Leaf and Yard Waste Composting Collection." Accessed Nov. 21, 2012: http://www.nyc.gov/html/nycwasteless/html/compost/collections_leaves.shtml.

City of New York Department of Transportation. 2013. "Questions and Answers to Common Sidewalk Violations." Accessed Aug. 19, 2013: http://www.nyc.gov/html/dot/html/infrastructure/sidewalk-faq.shtml.

City of Toronto. 2000. *Clean, Green, and Healthy: A Plan for an Environmentally Sustainable Toronto.* Environmental Task Force Final Report, February 2000. Accessed Mar. 11, 2011: http://www.toronto.ca/council/etfepfin.pdf.

Clancy, Kate. 2004. "Potential Contributions of Planning to Community Food Systems." *Journal of Planning Education and Research* 22: 435–38. doi: 10.1177/0739456X04264893.

Cohen, Nevin. 2011. "How Great Cities Are Fed Revisited: 10 Municipal Policies to Support the New York City Foodshed." *Fordham Environmental Law Journal* 42 (3): 691–710.

Cohen, Nevin, Kristin Reynolds, and Rupal Sanghvi. 2012. *Five Borough Farm: Seeding the Future of Urban Agriculture in New York City.* Design Trust for Public Space and Added Value (July 2012).

Connolly, James J., Erika S. Svendsen, Dana R. Fisher, and Lindsay K. Campbell. 2014. "Networked Governance and the Management of Ecosystem Services: The Case of Urban Environmental Stewardship in New York City." *Ecosystem Services* 10: 187–94.

——. 2013. "Organizing Urban Ecosystem Services through Environmental Stewardship Governance in New York City." *Landscape and Urban Planning* 109: 76–84. doi:10.1016/j.landurbplan.2012.07.001.

Conrad, Michael, Kubi Ackerman, and Dimitra Gavrilak. 2011. "Infrastructure → Health: Modeling Production, Processing, and Distribution Infrastructure for a Resilient Regional Food System." Columbia University Urban Design Lab. Accessed Aug. 19, 2013: http://www.urbandesignlab.columbia.edu/sitefiles/file/optimization-model.pdf.

Cornwell, B., and J. A. Harrison. 2004. "Union Members and Voluntary Association." *American Sociological Review* 69: 862–81. doi:10.1177/000312240406900606.

Cranz, G. 1982. *The Politics of Park Design, A History of Urban Parks in America.* Cambridge: MIT Press.

Crehan, Kate. 2002. *Gramsci, Culture, and Anthropology.* Berkeley: University of California Press.

Creswell, J. 1997. *Research Design: Qualitative, Quantitative, and Mixed Method Approaches.* Thousand Oaks, CA: Sage Publications.

Crona, B., and O. Bodin. 2006. "What You Know Is Who You Know? Communication Patterns among Resources Users as a Prerequisite for Comanagement." *Ecology and Society* 11 (2). http://www.ecologyandsociety.org/vol11/iss2/art7.

Davidson, D. J., and S. Frickel. 2004. "Understanding Environmental Governance—A Critical Review." *Organization & Environment* 17 (4): 471–92. doi:10.1177/1086026603259086.

Davies, Jonathan. 2011. *Challenging Governance Theory: From Networks to Hegemony.* Bristol: The Policy Press.

——. 2005. "Local Governance and the Dialectics of Hierarchy, Market and Network." *Policy Studies* 26 (3,4): 311–35. doi:10.1080/01442870500198379.

Demeritt, David. 2002. "What Is the 'Social Construction of Nature'? A Typology and Sympathetic Critique." *Progress in Human Geography* 26 (6): 767–90. doi:10.1191/0309132502ph402oa.

Desrocher, Pierre, and Hiroko Shimizu. 2012. *The Locavore's Dilemma: In Praise of the 10,000 Mile Diet.* New York: Public Affairs.

Dey, I. 1999. *Grounding Grounded Theory: Guidelines for Qualitative Inquiry*. San Diego, CA: Academic Press.

Dolnick, Sam. 2010. "The Hunger-Obesity Paradox." *New York Times,* Mar. 12, 2012. Accessed Nov. 29, 2012: http://www.nytimes.com/2010/03/14/nyregion/14hunger.html.

Donovan, G. H., and D. T. Butry. 2010. "Trees in the City: Valuing Street Trees in Portland, Oregon." *Landscape and Urban Planning* 94 (2): 77–83. doi:10.1016/j.landurbplan.2009.07.019.

Dowling, Robin. 2005. "Power, Subjectivity and Ethics in Qualitative Research." In *Qualitative Methods in Human Geography* 2nd ed., edited by Iain Hay, 19–29. Oxford: Oxford University Press.

Dryzek, John. 1990. *Discursive Democracy: Politics, Policy, and Political Science.* New York: Cambridge University Press.

Dunn, Kevin. 2005. "Interviewing." In *Qualitative Methods in Human Geography* 2nd ed., edited by Iain Hay, 79–105. Oxford: Oxford University Press.

Dupuis, E. Melanie, and David Goodman. 2005. "Should We Go 'Home' to Eat?: Toward a Reflexive Politics of Localism." *Journal of Rural Studies* 21: 359–71. doi:10.1016/j.jrurstud.2005.05.011.

Edwards, Michael, and David Hulme. 2002. "NGO Performance and Accountability: Introduction and Overview." In *Earthscan Reader on NGO Management,* edited by Michael Edwards and Alan Fowler, 187–203. London: Earthscan.

Eichtenthal, David. 1990. "The Other Elected Officials." In *Urban Politics New York Style,* edited by Jewel Bellush and Dick Netzger, 86–106. Armonk, NY: M. E. Sharp, Inc.

Elkin, S. L. 1987. *City and Regime in the American Republic.* Chicago: University of Chicago Press.

Ernstson, H., S. Sorlin, and T. Elmqvist. 2008. "Social Movements and Ecosystem Services—The Role of Social Network Structure in Protecting and Managing Urban Green Areas in Stockholm." *Ecology and Society* 13 (2): http://www.ecologyandsociety.org/vol13/iss2/art39.

Falxa-Raymond, N., M. I. Palmer, T. McPhearson, and K. L. Griffin. 2014. "Foliar Nitrogen Characteristics of Four Tree Species Planted in New York City Forest Restoration Sites." *Urban Ecosystems* 17: 807–24. doi:10.1007/s11252-014-0346-3.

Falxa-Raymond, Nancy, Erika S. Svendsen, and Lindsay K. Campbell. 2013. "From Job Training to Green Jobs: A Case Study of a Young Adult Employment Program Centered on Environmental Restoration in New York City." *Urban Forestry & Urban Greening* 12: 287–95. doi:10.1016/j.ufug.2013.04.003.

Farming Concrete. 2011. "Farming Concrete 2011 Harvest Report." Accessed Nov. 28, 2012: http://farmingconcrete.org/wp-content/uploads/2012/05/2011report.pdf.

Ferman, Barbara. 1996. *Challenging the Growth Machine: Neighborhood Politics in Chicago and Pittsburgh.* Lawrence: University of Kansas Press.

Fermino, Jennifer. 2015. "Michael Bloomberg Joins Mayor de Blasio to Celebrate Planting More than 1 Million Trees in NYC over Past Eight Years." *Daily News,*

Nov. 20, 2015. Accessed Feb. 25, 2016: http://www.nydailynews.com/new-york/bloomberg-de-blasio-celebrate-planting-1-million-nyc-trees-article-1.2441635.

Ferris, J., C. Norman, and J. Sempik. 2001. "People, Land and Sustainability: Community Gardens and the Social Dimension of Sustainable Development." *Social Policy & Administration* 35 (5): 559–68. doi:10.1111/1467-9515.t01-1-00253.

Fields, Tanya. 2013. "[Food Justice] End the Corporate Exploitation of 'Food Deserts.'" *Ebony,* Apr. 3, 2013. Accessed Jan. 5, 2016: http://www.ebony.com/wellness-empowerment/food-justice-end-the-corporate-exploitation-of-food-deserts#axzz3wP9e4Ojn.

Filipovic, Jill. 2012. "The Painful Lessons of the Central Park Five and the Jogger Rape Case." *The Guardian,* Oct 5, 2012. Accessed June 13, 2013: http://www.guardian.co.uk/commentisfree/2012/oct/05/central-park-five-rape-case.

Finn, Donovan, and Lynne McCormick. 2011. "Urban Climate Change Plans: How Holistic?" *Local Environment* 16 (4): 397–416. doi:10.1080/13549839.2011.579091.

Fischer, F., and M.A. Hajer. 1999. *Living with Nature: Environmental Politics as Cultural Discourse.* Oxford: Oxford University Press.

Fisher, Dana R., Lindsay K. Campbell, and Erika S. Svendsen. 2012. "The Organisational Structure of Urban Environmental Stewardship." *Environmental Politics* 21 (1): 26–48. doi:10.1080/09644016.2011.643367.

Fisher, Dana L., James J. Connolly, Erika S. Svendsen, and Lindsay K. Campbell. 2011. "Digging Together: Why People Volunteer to Help Plant One Million Trees in New York City." Environmental Stewardship Project at the Center for Society and Environment of the University of Maryland White Paper #2. Accessed Nov. 13, 2012: http://nrs.fs.fed.us/nyc/local-resources/downloads/Digging_Together_White_Paper.pdf.

——. 2010. "Who Volunteers to Steward the Urban Forest in New York City? An Analysis of Participants in MillionTreesNYC Planting Events." Environmental Stewardship Project White Paper #1. Accessed Nov. 13, 2012: http://nrs.fs.fed.us/nyc/local-resources/downloads/2010_WhoAreVolunteerStewards_ESP-Whitepaper1.pdf.

Fisher, Dana R., Erika S. Svendsen, and James Connolly. 2015. *Urban Environmental Stewardship and Civic Engagement: How Planting Trees Strengthens the Roots of Democracy.* London and New York: Routledge.

596 Acres. 2012. "596 Acres." Accessed Nov. 27, 2012: http://596acres.org/.

Flyvbjerg, B. 2001. *Making Social Science Matter: Why Social Inquiry Fails and How It Can Succeed Again.* Cambridge: Cambridge University Press.

Foderero, Lisa. 2014a. "Former Queens Fort Revamped for Work on Ecology." *New York Times,* Mar. 27, 2014. Accessed Aug. 28, 2014: http://www.nytimes.com/2014/03/28/nyregion/former-fort-revamped-for-work-on-ecology.html?hpw&rref=science&_r=1.

——. 2014b. "Focusing on Lesser-Known Open Spaces in New York." *New York Times,* May 6, 2014. Accessed Aug. 28, 2014: http://www.nytimes.com/2014/05/07/nyregion/focusing-on-lesser-known-open-spaces-in-new-york.html.

——. 2011. "As City Plants Trees, Some Say a Million Are Too Many." *New York Times,* Oct. 18, 2011. Accessed Oct. 19, 2011: http://www.nytimes.com/2011/10/19/nyregion/new-york-planting-a-million-treestoo-many-some-say.html?pagewanted=all.

Foglesong, R.E. 1986. *Planning the Capitalist City, The Colonial Era to the 1920s.* Princeton, NJ: Princeton University Press.

Food Systems Network New York City. 2010. "Food for the Future." Policy memo sent to the Mayor's Office of Long-Term Planning and Sustainability on Dec. 2, 2010. Accessed Aug. 19, 2013: http://www.foodsystemsnyc.org/files/FSNYC_2010_Food%20for%20the%20Future.pdf.

Foucault, Michel. 1991. "Governmentality." In *The Foucault Effect: Studies in Governmentality,* edited by G. Burchell, C. Gordon, and P. Miller. Chicago: Chicago University Press.

Fox, T., I. Koeppel, and S. Kellam. 1985. "Struggle for Space: The Greening of New York City, 1970–1984." NYC: Neighborhood Open Space Coalition.

Francis, Mark, Lisa Cashdan, and Lynn Paxson. 1984. *Community Open Spaces: Greening Neighborhoods through Community Action and Land Conservation.* Washington, D.C.: Island Press.

Freedlander, David. 2010. "Mrs. Mayor: Patti Harris Is the Gatekeeper for a Bloomberg Empire That Extends Well beyond Government." *New York Observer,* Dec. 8, 2012. Accessed Dec. 8, 2011: http://observer.com/2010/12/mrs-mayor-patti-harris-is-the-gatekeeper-for-a-bloomberg-empire-that-extends-well-beyond-government/.

Fuchs, Esther. 2011. "New York City as a Sustainable City." Columbia University Earth Institute Panel, Apr. 7, 2011.

Fusco, D. 2001. "Creating Relevant Science through Urban Planning and Gardening." *Journal of Research in Science Teaching* 38 (8): 860–77. doi:10.1002/tea.1036.

Gabriel, N. 2011. "The Work That Parks Do: Toward an Urban Environmentality." *Social and Cultural Geography* 12 (2): 123–41. doi:10.1080/14649365.2011.545139.

Gandy, M. 2002. *Concrete and Clay.* Cambridge, MA: MIT Press.

Getches, David, and David Pellow. 2002. "Beyond 'Traditional' Environmental Justice." In *Justice and Natural Resources: Concepts, Strategies, and Applications,* edited by Kathryn Mutz, Gary C. Bryner, and Douglas S. Kenney. chapter 1. Washington, D.C.: Island Press.

Gibbs, D., and A. E. G. Jonas. 2000. "Governance and Regulation in Local Environmental Policy: The Utility of a Regime Approach." *Geoforum* 31 (3): 299–313. doi:10.1016/S0016-7185(99)00052-4.

Gibbs, D., and R. Krueger. 2007. "Containing the Contradictions of Rapid Development? New Economy Spaces and Sustainable Urban Development." In *The Sustainable Development Paradox: Urban Political Economy in the United States and Europe,* edited by R. Krueger and D. Gibbs, 95–122. New York: Guilford Press.

Gittleman, Mara, Kelli Jordan, and Eric Brelsford. 2012. "Using Citizen Science to Quantify Community Garden Crop Yields." *Cities and the Environment* 5 (1): article 4: http://digitalcommons.lmu.edu/cate/vol5/iss1/4.

Glaeser, Edward, and Matthew Kahn. 2008. "The Greenness of Cities." Harvard University John F. Kennedy School of Government. Rappaport Institute for Greater Boston, Taubman Center for State and Local Government. Policy Brief, Mar. 2008. Accessed Nov. 28, 2012: http://www.hks.harvard.edu/var/ezp_site/storage/fckeditor/file/pdfs/centers-programs/centers/taubman/policybriefs/greencities_final.pdf.

Glover, T. D. 2003. "The Story of the Queen Anne Memorial Garden: Resisting a Dominant Cultural Narrative." *Journal of Leisure Research* 35 (2): 190–212.

Glover, T. D., K. J. Shinew, and D. C. Parry. 2005. "Association, Sociability, and Civic Culture: The Democratic Effect of Community Gardening." *Leisure Sciences* 27 (1): 75–92. doi:10.1080/01490400590886060.

Gottleib, Robert. 2010. "The Right to Place: Food, Streets and Immigrants." Pat Brown Institute: 1–6. Accessed Aug. 21, 2013: http://scholar.oxy.edu/cgi/viewcontent.cgi?article=1424&context=uep_faculty.

Greenhorns. 2012. "FAQs." Accessed Nov. 26, 2012: http://www.thegreenhorns.net/?cat=20#.

GreenThumb: New York City Department of Parks and Recreation. 2011. "GreenThumb Gardener's Handbook." Accessed Nov. 20, 2012: http://greenthumbnyc.org/pdf/gardeners_handbook.pdf.

——. 2010. "Welcome to GreenThumb" Accessed Dec. 16, 2010: http://www.greenthumbnyc.org/.

Grogan, Paul S., and Tony Proscio. 2000. *Comeback Cities: A Blueprint for Urban Neighborhood Revival.* Boulder: Westview Press.

Grove, J. Morgan, Jarlath O'Neil-Dunne, Keith Pelletier, David Nowak, and Jeff Walton. 2005. "A Report on New York City's Present and Possible Urban Tree Canopy." U.S. Department of Agriculture, Forest Service Northeastern Research Station. Report Prepared for Fiona Watt, Chief Forestry & Horticulture, Department of Parks & Recreation, City of New York. Accessed Oct. 22, 2012: http://nrs.fs.fed.us/nyc/local-resources/downloads/Grove_UTC_NYC_FINAL.pdf.

Grow to Learn. 2013. "Summer Cover Crops." Accessed May 15, 2013: http://growtolearn.org/pdf/Let_It_Be.pdf.

Growing Food and Justice for All Initiative. 2012. "About Us." Accessed Dec. 5, 2012: http://www.growingfoodandjustice.org/About_Us.html.

GrowNYC. 2012. "FAQs." Accessed Nov. 26, 2012: http://www.grownyc.org/about/faq.

——. 2011. "Farmers on the Edge: An Assessment of Greenmarket Farmers' Needs, and the Growing Challenges of Keeping Their Farms Viable." Accessed Nov. 26, 2012: http://www.grownyc.org/files/gmkt/farmers_on_the_edge.pdf.

——. 2010. "Greenmarket Farmers Markets" Accessed Dec. 16, 2010: http://www.grownyc.org/greenmarket.

Grynbaum, Michael. 2012. "Bloomberg Plans to Ban Sale of Big Sizes of Sugary Drinks." *New York Times,* May 30, 2012. Accessed Nov. 27, 2012: http://www.nytimes.com/2012/05/31/nyregion/bloomberg-plans-a-ban-on-large-sugared-drinks.html?pagewanted=all&_r=0.

Gussow, Joan Dye. 2006. "Reflections on Nutritional Health and the Environment: The Journey to Sustainability." *Journal of Hunger & Environmental Nutrition* 1: 1, 3–25. doi:10.1300/J477v01n01_02.

——. 1999. "Dietary Guidelines for Sustainability: Twelve Years Later." *Journal of Nutrition Education* July/August. doi:10.1016/S0022-3182(99)70441-3.

Gussow, Joan Dye, and Katherine Clancy. 1986. "Dietary Guidelines for Sustainability." *Journal of Nutrition Education* 18 (1): 1–5. doi:10.1016/S0022-3182(86)80255-2.

Gustavsson, E., I. Elander, and M. Lundmark. 2009. "Multilevel Governance, Networking Cities, and the Geography of Climate-change Mitigation: Two Swedish Examples." *Environment and Planning C: Government and Policy* 27 (1): 59–74. doi:10.1068/c07109j.

Guthman, Julie. 2011. *Weighing In: Obesity, Food Justice, and the Limits of Capitalism.* Berkeley: University of California Press.

Hajer, M. 1995. *The Politics of Environmental Discourse: Ecological Modernization and the Policy Process.* Oxford: Oxford University Press.

Hanneman, Robert A., and Mark Riddle. 2005. *Introduction to Social Network Methods.* Riverside: University of California, Riverside. Accessed June 6, 2013: http://faculty.ucr.edu/~hanneman/.

Haraway, D. 1991. "Situated Knowledges: The Science Question in Feminism and the Privilege of Partial Perspective." In Haraway, *Simians, Cyborgs, and Women: The Reinvention of Nature*, 183–201. New York: Routledge.

Harden, E. 1999. "Neighbors Give Central Park a Wealthy Glow." *New York Times*, Nov. 22, 1999. Accessed Dec. 13, 2009: http://www.nytimes.com/1999/11/22/nyregion/neighbors-give-central-park-a-wealthy-glow.html?pagewanted=1.

Hartig, T., M. Mang, and G. W. Evans. 1991. "Restorative Effects of Natural Environment." *Environment and Behavior* 23: 3–26. doi: 10.1177/0013916591231001.

Harvey, D. 2005. *A Brief History of Neoliberalism*. Oxford: Oxford University Press.

——. 1996. *Justice, Nature and the Geography of Difference.* Cambridge, MA: Blackwell Publishers Inc.

——. 1989. "From Managerialism to Entrepreneurialism: The Transformation in Urban Governance in Late Capitalism." *Geografiska Annaler* 71 (1): 3–17. doi:10.2307/490503.

Haub, Carl. 2012. "Changing the Way US Hispanics Are Counted." Population Reference Bureau. Accessed June 10, 2013: http://www.prb.org/Articles/2012/us-census-and-hispanics.aspx.

Healey, Patsy. 1997. *Collaborative Planning: Shaping Places in Fragmented Societies.* Vancouver: University of British Columbia Press.

Heehs, Jeff. 2008. "Reflections on the First Annual Gathering of the Growing Food and Justice for All Initiative." Food Systems Network NYC. Oct. 9, 2008. Accessed Dec. 5, 2012: http://www.foodsystemsnyc.org/article/reflections-first-annual-gathering-growing-food-and-justice-all-initiative.

Heynen, N. 2003. "The Scalar Production of Injustice in the Urban Forest." *Antipode* 35 (5): 980–98. doi:10.1111/j.1467-8330.2003.00367.x.

Henyen, Nik, Maria Kaika, and Erik Swyngedouw. 2006. "Urban Political Ecology: Politicizing the Production of Urban Natures." In *In the Nature of Cities: Urban Political Ecology and the Politics of Urban Metabolism*, edited by N. Heynen, M. Kaika, and E. Swyngedouw, 1–20. London: Routledge.

Heynen, Nik, and Harold Perkins. 2007. "Scalar Dialectics in Green: Urban Private Property and the Contradictions of the Neoliberalization of Nature." In *Neoliberal Environments: False Promises and Unnatural Consequences,* edited by Nik Heynen, James McCarthy, Scott Prudham, and Paul Robbins, 190–201. London: Routledge.

Hinrichs, C. Clare. 2003. "The Practice and Politics of Food System Localization." *Journal of Rural Studies* 19: 33–45. doi:10.1016/S0743-0167(02)00040-2.

Hodgson, Kimberly. 2012. "Planning for Food Access and Community-Based Food Systems: A National Scan and Evaluation of Local Comprehensive and Sustainability Plans." American Planning Association report. Accessed Dec. 20, 2012: http://www.planning.org/research/foodaccess/pdf/foodaccessreport.pdf.

Hoffman Brandt, Denise. 2012. *City Sink: Carbon Cycle Infrastructure for Our Built Environments*. New York: Oscar Riera Ojeda Pres.

Huber, Bridget. 2011. "Walmart's Fresh Food Makeover." *The Nation*, Sept. 14, 2011. Accessed Jan. 5, 2016: http://www.thenation.com/article/walmarts-fresh-food-makeover/.

International Council for Local Environmental Initiatives. 2010a. "About ICLEI." Accessed May 9, 2010: http://www.iclei.org/index.php?id=about.

——. 2010b. "The Process Behind PlaNYC: How the City of New York Developed Its Comprehensive Long-Term Sustainability Plan." Apr. 2010. Accessed Sept. 1, 2010: www.nyc.gov/html/planyc2030/downloads/pdf/iclei_planyc_case_study_201004.pdf.

ioby. 2011. "ioby Giving Report 2011." Accessed Nov. 27, 2012: https://ioby.org/files/uploads/2011%20Giving%20Report%20Final%20121411.pdf.
Jackson, Kenneth. 1985. *Crabgrass Frontier: The Suburbanization of the United States.* Oxford: Oxford University Press.
Jackson, Laura E. 2003. "The Relationship of Urban Design to Human Health and Condition." *Landscape and Urban Planning* 64 (4): 191–200. doi:10.1016/S0169-2046(02)00230-X.
Jessop, Bob. 2002. *The Future of the Capitalist City.* London: Polity.
——. 1994. "Post-Fordism and the State." In *Post-Fordism: A Reader,* edited by A. Amin, 251–79. Oxford: Blackwell.
Johnson, Lea. 2012. "Long-Term Outcomes of Urban Forest Restoration in New York City." New York City Urban Field Station Seminar series. Jan. 18, 2012.
Jonas, A. E. G., and D. C. Gibbs. 2003. "Changing Local Modes of Economic and Environmental Governance in England: A Tale of Two Areas." *Social Science Quarterly* 84: 1018–37. doi:10.1046/j.0038-4941.2003.08404011.x.
Jonas, A. E. G., and A. While. 2007. "Greening the Entrepreneurial City?: Looking for Spaces of Sustainability Politics in the Competitive City." In *The Sustainable Development Paradox: Urban Political Economy in the United States and Europe,* edited by R. Krueger and D. Gibbs, 123–59. New York: Guilford Press.
Jonas, Andrew, and David Wilson. 1999. *The Urban Growth Machine: Critical Perspectives Two Decades Later.* Albany: SUNY Press.
Jordan, A. 2008. "The Governance of Sustainable Development: Taking Stock and Looking Forwards." *Environment and Planning C: Government and Policy* 26 (1): 17–33. doi:10.1068/cav6.
Just Food. 2016. "CSA in NYC." Accessed Oct. 27, 2016: http://www.justfood.org/csa/csa-in-nyc.
Kaika, M. 2005. *City of Flows: Modernity, Nature and the City.* London and New York: Routledge.
Keck, Margaret E., and Kathryn Sikkink. 1998. *Activists beyond Borders: Advocacy Networks in International Politics.* Ithaca, NY: Cornell University Press.
Keil, Roger, and Julie-Anne Boudreau. 2006. "Metropolics and Metabolics: Rolling Out Environmentalism in Toronto." In *In the Nature of Cities: Urban Political Ecology and the Politics of Urban Metabolism,* edited by N. Heynen, M. Kaika, and E. Swyngedouw, 41–62. London: Routledge.
Kingsley, Jonathan "Yotti," Mardie Townsend, and Claire Henderson-Wilson. 2009. "Cultivating Health and Wellbeing: Members' Perceptions of the Health Benefits of a Port Melbourne Community Garden." *Leisure Studies* 28 (2): 207–19. doi:10.1080/02614360902769894.
Kjaer, Anne Mette. 2009. "Governance and the Urban Bureaucracy." In *Theories of Urban Politics* 2nd ed., edited by Jonathan Davies and David Imbroscio, 137–52. Los Angeles: Sage Publications.
Kliff, Sarah. 2012. "Mayor Mike Bloomberg, Public Health Autocrat: A Brief History." *Wonkblog, The Washington Post,* June 4, 2012. Accessed Nov. 27, 2012: http://www.washingtonpost.com/blogs/ezra-klein/post/mayor-mike-bloomberg-public-health-autocrat-a-brief-history/2012/06/04/gJQArSJbDV_blog.html.
Kneafsey, Moya. 2010. "The Region in Food—Important or Irrelevant?" *Cambridge Journal of Regions, Economy and Society* 3: 177–90. doi:10.1093/cjres/rsq012.
Knight-Ridder Newspapers. 1989. "Central Park Gang Rape Stirs Racial Tension." *Chicago Tribune,* Apr. 26, 1989. Accessed June 12, 2013: http://articles.chicagotribune.com/1989-04-26/news/8904070497_1_teens-face-charges-central-park-reservoir-joggers.
Kooiman, J. 1993. *Modern Governance.* Newbury Park, CA: Sage.

Krieger, James W., Tim K. Takaro, Lin Song, and Marcia Weaver. 2005. "The Seattle-King County Healthy Homes Project: A Randomized, Controlled Trial of a Community Health Worker Intervention to Decrease Exposure to Indoor Asthma Triggers." *American Journal of Public Health* 95 (4): 652–59. doi: 10.2105/AJPH.2004.042994.

Krueger, R., and J. Agyeman. 2005. "Sustainability Schizophrenia or 'Actually Existing Sustainabilities'? Toward a Broader Understanding of the Politics and Promise of Local Sustainability in the US." *Geoforum* 36: 410–17. doi:10.1016/j.geoforum.2004.07.005.

Kuo, F. E. 2001. "Coping with Poverty: Impacts of Environment and Attention in the Inner City." *Environment and Behavior* 33 (1): 5–34. doi:10.1177/00139160121972846.

Kuo, F. E., and W. C. Sullivan. 2001. "Aggression and Violence in the Inner City: Effects of Environment via Mental Fatigue." *Environment and Behavior* 33 (4): 543–71.

Kweon, B. S., W. C. Sullivan, and A. Wiley. 1998. "Green Common Spaces and the Social Integration of Inner-City Older Adults." *Environment and Behavior* 30 (6): 832–58. doi: 10.1177/00139160121973124.

Lake, R. 2003. "Dilemmas of Environmental Planning in Post-Urban New Jersey." *Social Science Quarterly* 84: 1003–17. doi:10.1046/j.0038-4941.2003.08404013.x.

——. 1996. "Volunteers, NIMBYs and Environmental Justice: Dilemmas of Democratic Practice." *Antipode* 28 (2): 160–74. doi:10.1111/j.1467-8330.1996.tb00520.x.

Lander, Brad. 2012. "How the NYC Government Can Impact Our Food System." Panel Session, Brooklyn Food Conference, May 12, 2012.

Latour, B. 2005. *Reassembling the Social: An Introduction to Actor-Network-Theory*. Oxford: Oxford University Press.

Lauria, M., 1997. "Introduction." In *Reconstructing Urban Regime Theory: Regulating Politics in the Global Economy*, edited by M. Lauria, 1–110. London: Sage.

Lawson, Laura. 2005. *City Bountiful: A Century of Community Gardening in America*. Berkeley: University of California Press.

Lees, Loretta. 2004. "Urban Geography: Discourse Analysis and Urban Research." *Progress in Human Geography* 28 (1): 101–7. doi:10.1191/0309132504ph473pr.

Lefebvre, Henri. 1991. *The Production of Space*. Oxford: Blackwell.

Lefer, David. 1999. "Gardens Flap Growing Giuliani to Auction 126 Plots." *New York Daily News*, Feb. 21, 1999. Accessed June 12, 2013: http://www.nydailynews.com/archives/news/gardens-flap-growing-giuliani-auction-126-plots-article-1.835298.

Lemire, Jonathan. 2015. "De Blasio and Bloomberg make Peace at NYC Tree-Planting." *Washington Times*, Nov. 20, 2015. Accessed Feb. 25, 2016: http://www.washingtontimes.com/news/2015/nov/20/de-blasio-and-bloomberg-make-peace-at-tree-planting/.

Levin, Stephen. 2012. "How the NYC Government Can Impact Our Food System." Panel Session, Brooklyn Food Conference, May 12, 2012.

Light, A. 2003. "Urban Ecological Citizenship." *Journal of Social Philosophy* 34 (1): 44–63. doi:10.1111/1467-9833.00164.

Locke, D. H., J. M. Grove, M. F. Galvin, J. P. M. O'Neil-Dunne, and C. Murphy. 2013. "Applications of Urban Tree Canopy Assessment and Prioritization Tools: Supporting Collaborative Decision Making to Achieve Urban Sustainability Goals." *Cities and the Environment (CATE)* 6 (1): http://digitalcommons.lmu.edu/cate/vol6/iss1/7/.

Locke, Dexter H., Kristen L. King, Erika S. Svendsen, Lindsay K. Campbell, Christopher Small, Nancy F. Sonti, Dana R. Fisher, and Jacqueline W. T. Lu. 2014. "Urban

Environmental Stewardship and Changes in Vegetative Cover and Building Footprint in New York City Neighborhoods (2000–2010)." *Journal of Environmental Studies and Sciences* 4 (3): 250–62. doi:10.1007/s13412-014-0176-x.

Loftus, A., and F. Lumsden. 2008. "Reworking Hegemony in the Urban Waterscape." *Transactions of the Institute of British Geographers* 33: 109–26. doi:10.1111/j.1475-5661.2007.00280.x.

Logan, John, and Harvey Molotch. 1987. *Urban Fortunes: The Political Economy of Place.* Berkeley: University of California Press.

Logan, J., R. Whaley, and K. Crowder. 1997. "The Character and Consequences of Growth Regimes: An Assessment of 20 Years of Research." *Urban Affairs Quarterly* 32 (5): 603–30. doi:10.1177/107808749703200501.

Lorimer, J. 2006. "What about the Nematodes? Taxonomic Partialities in the Scope of UK Biodiversity Conservation." *Social & Cultural Geography* 7 (4): 539–58. doi:10.1080/14649360600825687.

Lovasi, Gina S., James W. Quinn, Kathryn M. Neckerman, Matthew S. Perzanowski, and Andrew Rundle. 2008. "Children Living in Areas with More Street Trees Have Lower Asthma Prevalence." *J. Epidemiology and Community Health* (published online May 1, 2008). doi:10.1136/jech.2007.071894.

Lu, Jacqueline W. T., Erika S. Svendsen, Lindsay K. Campbell, Jennifer Greenfeld, Jessie Braden, and Kristy King. 2011. "Biological, Social, and Neighborhood Design Factors in Young Street Tree Mortality in New York City." *Cities and the Environment* 3 (1): 1–15. http://digitalcommons.lmu.edu/cate/vol3/iss1/5/.

Lukes, Stephen. 2005. *Power: A Radical View* 2nd ed. New York: Palgrave Macmillan.

Luley, C. J., and J. Bond. 2002. A Report to the Northeast State Foresters Association. A Plan to Integrate Management of Urban Trees into Air Quality Planning. Naples, NY: Davey Resources Group.

Lyson, Thomas A. 2007. "Civic Agriculture and the North American Food System." In *Remaking the North American Food System: Strategies for Sustainability,* edited by C. Clare Hinrichs and Thomas A. Lyson, 19–32. Lincoln: University of Nebraska Press.

Machlis, Gary. 2015. "Multi-Level Solutions for Health." Panel presentation at the *Urban Nature as a Health Resource Conference,* Yale University, New Haven, CT, Feb. 6, 2015.

Maddox, David, comp. ed. 2010. "Supporting Success: Making the Transition to Green Collar Jobs." New York, NY: MillionTreesNYC, unpublished report. Accessed Nov. 14, 2012: http://nrs.fs.fed.us/nyc/local-resources/downloads/Supporting_Success_Workshop_Report_Summer_2010.pdf.

Mandelbaum, S. 2007. "Review of PlaNYC: A Greener, Greater New York." *Journal of Planning Education and Research* 27 (2): 231. doi:10.1177/0739456X07308627.

Marsden, P. 1990. "Network Data and Measurement." *Annual Review of Sociology* 16: 435–63. doi:10.1146/annurev.so.16.080190.002251.

Martin, Deborah. 2004. "Nonprofit Foundations and Grassroots Organizing: Reshaping Urban Governance." *The Professional Geographer* 56 (3): 394–405. doi:10.1111/j.0033-0124.2004.05603008.x.

Marvy, Ian. 2009. "Youth Empowerment through Urban Agriculture: Red Hook Community Farm." In *Restorative Commons: Creating Health and Well-being through Urban Landscapes,* General Technical Report, edited by Lindsay Campbell and Anne Wiesen, 202–15. Newtown Square, PA: U.S. Department of Agriculture, Forest Service, Northern Research Station.

McCarthy, J. 2002. "First World Political Ecology: Lessons from the Wise Use Movement." *Environment and Planning A* 34: 1281–302. doi:10.1068/a3526.

McCarthy, James, and Scott Prudham. 2004. "Neoliberal Nature and the Nature of Neoliberalism." *Geoforum* 35: 275–83. doi:10.1016/j.geoforum.2003.07.003.

McGranahan, Gordon, Peter Marcotullio, Xuemi Bai, Deborah Balk, Tania Braga, Ian Douglas, Thomas Elmqvist, William Rees, David Satterthwaite, Jacob Songsore, Hania Zlotnik, Jerry Eades, Exequiel Ezcurra. 2005. "Urban Systems." In *Millennium Ecosystem Assessment: Ecosystems and Human Well-Being*. Edited by Hassan Rashid, Scholes Robert, and Ash Neville, 795–825. Washington, D.C.: Island Press.

McMillan, Tracie. 2010. "An Urban Farming Pioneer Sows His Own Legacy." *New York Times*, May 18, 2010. Accessed Dec. 11, 2012: http://www.nytimes.com/2010/05/19/dining/19farm.html?pagewanted=all&_r=0.

McPhearson, P. T., M. Feller, A. Felson, R. Karty, J. W. T. Lu, M. I. Palmer, and T. Wenskus. 2010. "Assessing the Effects of the Urban Forest Restoration Effort of MillionTreesNYC on the Structure and Functioning of the New York City Ecosystems." *Cities and the Environment* 3 (1): article 7. http://digitalcommons.lmu.edu/cate/vol3/iss1/7/.

McPherson, E. Gregory, and James R. Simpson. 2002. "A Comparison of Municipal Forest Benefits and Costs in Modesto and Santa Monica, California, USA." *Urban Forestry and Urban Greening* 1: 61–74. doi:10.1078/1618-8667-00007.

McPherson, E. G., J. R. Simpson, P. J. Peper, and E. Aguaraon. 2008. "Urban Forestry and Climate Change." Albany, CA: U.S. Department of Agriculture, Forest Service, Pacific Southwest Research Station. Accessed June 3, 2013: http://www.fs.fed.us/ccrc/topics/urban-forests/.

McWilliams, James. 2009. *Just Food: Where Locavores Get It Wrong and How We Can Truly Eat Responsibly.* New York: Little, Brown, and Co.

Mees, Carolin, and Edie Stone. 2012. "Zoned Out: The Potential of Urban Agriculture Planning to Turn Against Its Roots." *Cities and the Environment* 5 (1): article 7. http://digitalcommons.lmu.edu/cate/vol5/iss1/7/.

Mele, C. 2000. "The Materiality of Urban Discourse." *Urban Affairs Review* 35: 628–48. doi:10.1177/10780870022184570.

Mertig, A. G., R. E. Dunlap, and D. E. Morrison. 2001. "The Environmental Movement in the United States." In *Handbook of Environmental Sociology,* edited by R. E. Dunlap and W. Michelson, 448–81. Westport, CT: Greenwood.

Miles, M., and A. M. Huberman. 1994. *Qualitative Data Analysis*. Thousand Oaks, CA: Sage Publications.

MillionTreesNYC. 2012. "MillionTreesNYC Adopt-a-Tree." Accessed Nov. 13, 2012: http://stewardship.nycparks.org/.

——. 2010. "About MillionTreesNYC" webpage. Accessed Dec. 16, 2010: http://milliontreesnyc.org/html/about/about.shtml.

——. 2009. "Congrats 'Green' Graduates! City Celebrates Inaugural Class of Million TreesNYC Training Program." Press release, May 29, 2009. Accessed Nov. 13, 2012: http://milliontreesnyc.org/html/newsroom/pr_inaugural_class.shtml.

MillionTreesNYC Advisory Board Research and Evaluation Subcommittee. 2009. "MillionTreesNYC, Green Infrastructure, and Urban Ecology: Building a Research Agenda." Report from the May 2009 Research Workshop. Accessed Nov. 14, 2012: http://nrs.fs.fed.us/nyc/local-resources/downloads/MillionTrees WorkshopReport.pdf.

Mitchell, Timothy. 2002. *Rule of Experts: Egypt, Techno-Politics, Modernity*. Berkeley: University of California Press.

Mollenkopf, John Hull. 1992. "Political Inequality." In *Dual City: The Restructuring of New York*, edited by John Hull Mollenkopf and Manuel Castells, 333–58. New York: Russell Sage Foundation.

Mollenkopf, John Hull, and Manuel Castells. 1992. "Introduction." In *Dual City: The Restructuring of New York,* edited by John Hull Mollenkopf and Manuel Castells, 3–22. New York: Russell Sage Foundation.

Molotch, Harvey. 1976. "The City as a Growth Machine: Toward a Political Economy of Place." *American Journal of Sociology* 82 (2): 309–32.

Morales, D. J. 1980. "The Contribution of Trees to Residential Property Value." *Journal of Arboriculture* 6: 305–8.

Morgan, Kevin, and Roberta Sonnino. 2010. "The Urban Foodscape: World Cities and the New Food Equation." *Cambridge Journal of Regions, Economy and Society* 3: 209–24. doi:10.1093/cjres/rsq00.7.

Moskell, C., S. Broussard Allred, and G. Ferenz. 2010. "Examining Volunteer Motivations and Recruitment Strategies for Engagement in Urban Forestry." *Cities and the Environment.* 3 (1): article 9. http://digitalcommons.lmu.edu/cate/vol3/iss1/9/.

Mossberger, Karen. 2009. "Urban Regime Analysis." In *Theories of Urban Politics* 2nd ed., edited by Jonathan Davies and David Imbroscio, 40–54. Los Angeles: Sage Publications.

Mossberger, K., and G. Stoker. 2001. "The Evolution of Urban Regime Theory—The Challenge of Conceptualization." *Urban Affairs Review* 36 (6): 810–35. doi:10.1177/10780870122185109.

Murphy-Dunning, Colleen. 2009. "From Front Yards to Street Corners: Revitalizing Neighborhoods through Community-based Land Stewardship." In *Restorative Commons: Creating Health and Well-being through Urban Landscapes,* General Technical Report, edited by Lindsay Campbell and Anne Wiesen, 154–63. Newtown Square, PA: U.S. Department of Agriculture, Forest Service, Northern Research Station.

National Young Farmers' Coalition. 2012. "Our Mission, Vision, and Guiding Principles." Accessed Nov. 26, 2012: http://www.youngfarmers.org/about/our-work/.

Navarro, Mireya. 2012. "Push to Give Midsize Farms a Hub at a Bronx Market" *New York Times,* Apr. 24, 2012. Accessed Dec. 3, 2012: http://www.nytimes.com/2012/04/25/nyregion/push-to-give-midsize-new-york-farms-a-hub-at-the-hunts-point-market.html?pagewanted=all&_r=0.

——. 2009. "Mayor's Environmental Record: Grand Plans and Small Steps Forward." *New York Times,* Oct. 23, 2009.

Neighborhood Open Space Coalition. 2000. "Community Garden Update." No. 54—Apr. 4, 2000. Accessed Nov. 20, 2012: http://hpn.asu.edu/archives/2000-April/000492.html.

Nestle, Marion. 2010. *Safe Food: The Politics of Food Safety.* Berkeley: University of California Press.

——. 2005. *What to Eat.* New York: North Point Press.

New York City Community Gardening Coalition. 2010. "Where We Stand and How We Got Here." Accessed Nov. 28, 2012: http://nyccgc.org/about/history/.

New York City Council. 2010. *FoodWorks: A Vision to Improve NYC's Food System.* Nov. 2010. Accessed Mar. 1, 2011: http://council.nyc.gov/html/food/files/foodworks_fullreport_11_22_10.pdf.

New York City Housing Authority. 2015. *NextGeneration NYCHA.* Accessed Nov. 2, 2015: https://www1.nyc.gov/site/nycha/about/nextgen-nycha.page.

——. 2012. "About NYCHA." Accessed Nov. 13, 2012: http://www.nyc.gov/html/nycha/html/about/about.shtml.

——. 2011. "Greening Public Housing: New York City Housing Authority." Powerpoint presentation. Fall 2011.

New York City Independent Budget Office. 2013. "Understanding New York City's Budget: A Guide to the Capital Budget." Accessed Aug. 15, 2014: http://www.ibo.nyc.ny.us/iboreports/IBOCBG.pdf.

New York Sun Works. 2012. "The Greenhouse Project." Accessed Nov. 27, 2012: http://nysunworks.org/thegreenhouseproject.

Newman, L., and A. Dale. 2004. "Network Structure, Diversity, and Proactive Resilience Building: A Response to Tompkins and Adger." *Ecology and Society* 10 (1). http://www.ecologyandsociety.org/vol10/iss1/resp2.

Nowak, David, Kevin L. Civerolo, S. Trivikrama Rao, Gopal Sistla, Christopher J. Luley, and Daniel E. Crane. 2000. "A Modeling Study of the Impact of Urban Trees on Ozone." *Atmospheric Environment* 34 (10): 1601–13. doi:10.1016/S1352-2310(99)00394-5.

Nowak, D. J., D. E. Crane, and J. C. Stevens. 2006. "Air Pollution Removal by Urban Trees and Shrubs in the United States." *Urban Forestry and Urban Greening* 4: 115–23. doi:10.1016/j.ufug.2006.01.007.

Nowak, David J., Susan M. Stein, Paula B. Randler, Eric J. Greenfield, Sara J. Comas, Mary A. Carr, and Ralph J. Alig. 2010. *Sustaining America's Urban Trees and Forests: A Forests on the Edge Report.* Gen. Tech. Rep. NRS-62. Newtown Square, PA: U.S. Department of Agriculture, Forest Service, Northern Research Station. http://www.treesearch.fs.fed.us/pubs/35572.

Ogden, Cynthia, and Margaret Carroll. 2010. "Prevalence of Obesity among Children and Adolescents: United States, Trends 1963–1965 through 2007–2008." CDC NCHS E-Stat. Accessed Aug. 13, 2013: http://www.cdc.gov/nchs/data/hestat/obesity_child_07_08/obesity_child_07_08.htm.

Ohmer, Mary, Pamela Meadowcroft, Kate Freed, and Ericka Lewis. 2009. "Community Gardening and Community Development: Individual, Social and Community Benefits of a Community Conservation Program." *Journal of Community Practice* 17: 377–99. doi:10.1080/10705420903299961.

Orsega-Smith, E., A. J. Mowen, L. L. Payne, and G. Godbey. 2004. "The Interaction of Stress and Park Use on Psycho-physiological Health in Older Adults." *Journal of Leisure Research* 36 (2): 232–57.

Ostrom, Marcia Ruth. 2007. "Community Supported Agriculture as an Agent of Change: Is it Working?" In *Remaking the North American Food System: Strategies for Sustainability,* edited by C. Clare Hinrichs and Thomas A. Lyson, 99–120. Lincoln: University of Nebraska Press.

Owen, David. 2009. *Green Metropolis: Why Living Smaller, Living Closer, and Driving Less Are the Keys to Sustainability.* New York: Riverhead Books.

——. 2004. "Green Manhattan." *New Yorker,* Oct. 18, 2004, 111–23.

Painter, Joe. 1997. "Regulation, Regime, and Practice in Urban Politics." In *Reconstructing Urban Regime Theory: Regulating Politics in the Global Economy,* edited by M. Lauria, 122–44. London: Sage.

Park, Alice. 2012. "The New York City Soda Ban and a Brief History of Bloomberg's Nudges." *Time,* May 31, 2012. Accessed Nov. 27, 2012: http://healthland.time.com/2012/05/31/bloombergs-soda-ban-and-other-sweeping-health-measures-in-new-york-city/.

Parr, H. 2007. "Mental Health, Nature Work, and Social Inclusion." *Environment and Planning D-Society & Space* 25 (3): 537–61. doi:10.1068/d67j.

Pataki, Diane E., Margaret M. Carreiro, Jennifer Cherrier, Nancy E. Grulke, Viniece Jennings, Stephanie Pincetl, Richard V. Pouyat, Thomas H. Whitlow, and Wayne C. Zipperer. 2011. "Coupling Biogeochemical Cycles in Urban Environments: Ecosystem Services, Green Solutions, and Misconceptions." *Frontiers in Ecology and the Environment* 9 (1): 27–36. doi:10.1890/090220.

Peck, Jamie, and Nik Theodore. 2015. *Fast Policy: Experimental Statecraft at the Thresholds of Neoliberalism.* Minneapolis: University of Minnesota Press.

Peck, Jamie, and Adam Tickell. 2002. "Neoliberalizaing Space." In *Spaces of Neoliberalism: Urban Restructuring in North America and Western Europe,* edited by N. Brenner and N. Theodore, 33–57. Oxford: Blackwell.

Peper, Paula J., E. Gregory McPherson, James R. Simpson, Shelley L. Gardner, Kelaine E. Vargas, and Qingfu Xiao. 2007. "New York City, New York Municipal Forest Resource Analysis." Center for Urban Forest Research, U.S. Department of Agriculture Forest Service, Pacific Southwest Research Station. March 2007. http://www.treesearch.fs.fed.us/pubs/45969.

Perkins, H.A. 2009. "Out from the (Green) Shadow? Neoliberal Hegemony through the Market Logic of Shared Urban Environmental Governance." *Political Geography* 28 (7): 395–405. doi:10.1016/j.polgeo.2009.09.007.

——. 2007. "Ecologies of Actor-networks and (Non)social Labor within the Urban Political Economies of Nature." *Geoforum* 38: 1152–62. doi:10.1016/j.geoforum.2007.01.007.

Peterson, Paul. 1981. *City Limits.* Chicago: University of Chicago Press.

Pierce, Neal R., and Curtis W. Johnson. 2008. *Century of the City: No Time to Lose.* New York: Rockefeller Foundation.

Pincetl, S. 2010. "Implementing Municipal Tree Planting: Los Angeles Million-Tree Initiative." *Environmental Management* 45 (2): 227–38. doi:10.1007/s00267-009-9412-7.

——. 2003. "Nonprofits and Park Provision in Los Angeles: An Exploration of the Rise of Governance Approaches to the Provision of Local Services." *Social Science Quarterly* 84: 979–1001. doi:10.1046/j.0038-4941.2003.08404019.x.

Piven, Frances Fox, and Richard A. Cloward. 1979. *Poor People's Movements: Why They Succeed, How They Fail.* New York: Vintage Books Edition.

Polanyi, Karl. 1944 (2001). *The Great Transformation: The Political and Economic Origins of Our Time.* Boston: Beacon Press.

Pollan, Michael. 2006. *The Omnivore's Dilemma: A Natural History of Four Meals.* New York: Penguin Press.

Pothukuchi, Kameshwari, and Jerome Kaufman. 2000. "The Food System: A Stranger to the Planning Field." *Journal of the American Planning Association* 66 (2): 113–24. doi:10.1080/01944360008976093.

Power, E. R. 2005. "Human-Nature Relations in Suburban Gardens." *Australian Geographer* 36 (1): 39–53. doi:10.1080/00049180500050847.

Prell, C., K. Hubacek, and M. Reed. 2009. "Stakeholder Analysis and Social Network Analysis in Natural Resource Management." *Society and Natural Resources* 22: 501–18. doi:10.1080/08941920802199202.

Prytherch, David. 2002. "Selling the Eco-entrepreneurial City: Natural Wonders and Urban Stratagems in Tucson, Arizona." *Urban Geography* 23 (8): 771–93.

Rae, Ruth A., G. Simon, and J. Braden. 2010. "Public Reactions to New Street Tree Planting." *Cities and the Environment* 3 (1): article 10. http://digitalcommons.lmu.edu/cate/vol3/iss1/10/.

Randall, David K. 2007. "Maybe Only God Can Make a Tree, but Only People Can Put a Price on It." *New York Times,* Apr. 18, 2007. Accessed Jan. 2, 2013: http://www.nytimes.com/2007/04/18/nyregion/18trees.html?_r=0.

Reynolds, K., and N. Cohen. 2016. *Beyond the Kale: Urban Agriculture and Social Justice Activist in New York City.* Athens: University of Georgia Press.

Rhodes, R.A.W. 1996. "The New Governance: Governing without Government." *Political Studies* 44: 652–67. doi:10.1111/j.1467-9248.1996.tb01747.x.

Rivera, Ray. 2007. "A Shady Plan by the Mayor That's Likely to Be Popular." *New York Times,* Apr. 22, 2007.

Robbins, Paul. 2007. *Lawn People: How Grasses, Weeds, and Chemicals Make Us Who We Are.* Philadelphia: Temple University Press.

——. 2004. *Political Ecology: A Critical Introduction.* London: Blackwell.

Robbins, Paul, and Brian Marks. 2010. "Assemblage Geographies." In *Sage Handbook of Social Geographies,* edited by Susan Smith, Rachel Pain, Sallie Marston, and John Paul Jones III, 176–94. London: Sage.

Robbins, P., and J. Sharp. 2003. "The Lawn Chemical Economy and Its Discontents." *Antipode* 35 (5): 955–79. doi:10.1111/j.1467-8330.2003.00366.x.

Roberts, Sam. 2006. "Infamous 'Drop Dead' Was Never Said by Ford." *New York Times,* Dec. 28, 2006. Accessed June 12, 2013: http://www.nytimes.com/2006/12/28/nyregion/28veto.html?_r=0.

Rocheleau, D. 2008. "Political Ecology in the Key of Policy: From Chains of Explanation to Webs of Relation." *Geoforum* 39: 716–27. doi:10.1016/j.geoforum.2007.02.005.

Rocheleau, D., and R. Roth. 2007. "Rooted Networks, Relational Webs and Powers of Connection: Rethinking Human and Political Ecologies." *Geoforum* 38 (3): 433–37. doi:10.1016/j.geoforum.2006.10.003.

Roman, Lara. 2014. "How Many Trees Are Enough?" *Scenario Journal.* Accessed Aug. 28, 2014: http://scenariojournal.com/article/how-many-trees-are-enough/.

Romer, Nancy. 2012. "Closing Plenary." Brooklyn Food Conference, May 12, 2012.

Rosan, Christina. 2011. "Can PlaNYC Make New York City 'Greener and Greater' for Everyone?: Sustainability Planning and the Promise of Environmental Justice." *Local Environment* 17 (9): 959–76. doi:10.1080/13549839.2011.627322.

Rose, G. 2001. *Visual Methodologies: An Introdcution to the Interpretation of Visual Materials.* London: Sage Publications.

——. 1997. "Situating Knowledges: Positionality, Reflexivities and Other Tactics." *Progress in Human Geography* 21 (3): 305–20. doi: 10.1191/030913297673302122.

Rosen, Liza, and Jennifer Greenfeld. 2006. "Greening Morrisania: A Community Forestry Management Plan." City of New York Department of Parks and Recreation. June 2006. Accessed Jan. 2, 2013: www.nycgovparks.org/sub_your_park/trees_greenstreets/morrisania_forest_plan/Morrisania_Plan.pdf.

Rosenzweig, R. 1983. *Eight Hours for What We Will, Workers and Leisure in an Industrial City, 1970–1920.* Cambridge: Cambridge University Press.

Rubenstein, Dana. 2014. "Mayor Continues City Park Funding Policies—For Now." *Capital,* May 9, 2014. Accessed Aug. 28, 2014: http://www.capitalnewyork.com/article/city-hall/2014/05/8545118/mayor-continues-city-parks-funding-policies-now.

Ryan, Michael G., Mark E. Harmon, Richard A. Birdsey, Christian P. Giardina, Linda S. Heath, Richard A. Houghton, Robert B. Jackson, Duncan C. McKinley, James F. Morrison, Brian C. Murray, Diane E. Pataki, and Kenneth Skog. 2010. "A Synthesis of Science on Forests and Carbon for U.S. Forests." *Issues in Ecology* (Spring 2010). doi:10.1890/10-0697.1.

Ryzik, Melena. 2009. "Live, Art, and Chickens, Afloat in the Harbor." *New York Times,* Aug. 13, 2008. Accessed Dec. 3, 2010, via infoweb.newsbank.com.

Sadik, N. 1999. *The State of World Population 1999—6 Billion: A Time for Choices.* New York: United Nations Population Fund. Oct. 13, 2003. Accessed Mar. 1, 2011: www.unfpa.org/swp/1999/pdf/swp99.pdf.

Said, E. 2000. "Orientalism." In *The Edward Said Reader,* edited by M. Bayoumi and A. Ruben, 63–113. New York: Vintage Books.

Saldivar-Tanaka, L., and M. E. Krasny. 2004. "Culturing Community Development, Neighborhood Open Space, and Civic Agriculture: The Case of Latino Community Gardens in New York City." *Agriculture and Human Values* 21 (4): 399–412. doi:10.1007/s10460-003-1248-9.

Salkin, Allen. 2008. "Leaving behind the Trucker Hat." *New York Times,* Mar. 16, 2008. Accessed Dec. 3, 2010, via infoweb.newsbank.com.

Sandler, L., J. A. Vandergrift, and C. VerBrugghen. 1995. "From Desert to Garden: Reconnecting Disconnected Youth." *Educational Leadership* (May 1995): 14–16.

Sassen, Saskia. 2001. *The Global City: New York, London, Tokyo.* Princeton, NJ: Princeton University Press.

Savitch, H. V., and Paul Kantor. 2002. *Cities in the International Marketplace: The Political Economy of Urban Development in North America and Western Europe.* Princeton, NJ: Princeton University Press.

Saxon-Harrold, Susan. 1990. "Competition, Resources, and Strategy in the British Nonprofit Sector" In *The Third Sector: Comparative Studies in Nonprofit Organizations,* edited by Helmut K. Anheir and Wolfgang Seibel, 123–36. Berlin: de Gruyter.

Schlosberg, David. 2007. *Defining Environmental Justice: Theories, Movements and Nature.* Oxford: Oxford University Press.

——. 2003. "The Justice of Environmental Justice: Reconciling Equity, Recognition, and Participation in a Political Movement." In *Moral and Political Reasoning in Environmental Practice,* edited by A. Light and A. De-Shalit, 77–106. Cambridge, MA: MIT Press.

Schmelzkopf, K. 2002. "Incommensurability, Land Use, and the Right to Space: Community Gardens in New York City." *Urban Geography* 23 (4): 323–43. doi:10.2747/0272-3638.23.4.323.

——. 1995. "Urban Community Gardens as Contested Space." *Geographical Review* 85 (3): 364–81. doi:10.2307/215279.

Schwarz, K., M. Fragkias, C. G. Boone, W. Zhou, M. McHale, J. M. Grove, et al. 2015. "Trees Grow on Money: Urban Tree Canopy Cover and Environmental Justice." *PLoS ONE* 10 (4): e0122051. doi:10.1371/journal.pone.0122051.

Scobey, David. 2002. *Empire City: The Making and Meaning of the New York City Landscape.* Philadelphia: Temple University Press.

Scott, J. 2000. *Social Network Analysis: A Handbook* 2nd ed. London and Beverley Hills: Sage Publications.

Scott, James. 1998. *Seeing like a State: How Certain Schemes to Improve the Human Condition Have Failed.* New Haven: Yale University Press.

Secord, J., and A. Zakhor. 2007. "Tree Detection in Urban Regions Using Aerial Lidar and Image Data." *Geoscience and Remote Sensing Letters, IEEE* 4 (2): 196–200. doi:10.1109/LGRS.2006.888107.

Seeing Green. 2013. "Seeing Green." Accessed May 30, 2013: http://www.seeingreen.com/.

Segal, Paula. 2012. "Vacancy! Is There Enough Space for Food AND Housing in New York City?" Panel Session, Brooklyn Food Conference, May 12, 2012.

Severson, K. 2008. "Hungry for a Market, but Where?" *New York Times,* Jan. 2, 2008. Accessed Dec. 3, 2010, via infoweb.newsbank.com.

Shannon, J. 2013. "Food Deserts: Governing Obesity in the Neoliberal City." *Progress in Human Geography* 38 (2): 248–66. doi:10.1177/0309132513484378.

Sheffield, B. K. 1992. "The Affective and Cognitive Effects of an Interdisciplinary Garden-Based Curriculum on Underachieving Elementary Students." Ph.D. thesis, University of South Carolina College of Education, Columbia.

Shefter, Martin. 1985. *Political Crisis/Fiscal Crisis: The Collapse and Revival of New York City.* New York: Basic Books.

Sherrill, Sam. 2003. *Harvesting Urban Timber*. Fresno, CA: Linden Publishing.

Shute, Lindsey. 2011. "Building a Future with Farmers: Challenges Faced by Young, American Farmers and a National Strategy to Help Them Succeed." National Young Farmers Coalition report. Nov. 2011. Accessed Nov. 26, 2012: http://www.youngfarmers.org/reports/Building_A_Future_With_Farmers.pdf.

Simmons, B. L. 2016. Assessing Survival on MillionTreesNYC Reforestation Sites, Unpublished Data. New York City Department of Parks and Recreation.

——. 2010. "Assessing Survival on MillionTreesNYC Reforestation Sites." *Cities and the Environment* 3 (1): poster 22. http://digitalcommons.lmu.edu/cate/vol3/iss1/22/.

Simmons, Brady L., Richard A. Hallett, Nancy Falxa Sonti, D. S. N. Auyeung, and Jacqueline W. T. Lu. 2016. "Long-Term Outcomes of Forest Restoration in an Urban Park." *Restoration Ecology* 24 (1):109–18.

Singer, Peter, and Jim Mason. 2006. *The Ethics of What We Eat: Why Our Food Choices Matter*. New York: Rodale.

Sites, William. 2003. *Remaking New York: Primitive Globalization and the Politics of Urban Community.* Minneapolis: University of Minnesota Press.

——. 1997. "The Limits of Urban Regime Theory: New York City under Koch, Dinkins, and Giuliani." *Urban Affairs Review* 32 (4): 536–57. doi:10.1177/107808749703200405.

Skocpol, Theda. 1985. "Bringing the State Back In." In *Bringing the State Back In,* edited by Peter Evans, Dietrich Rueschemeyer, and Theda Skocpol, 3–43. Cambridge: Cambridge University Press.

Slow Food USA. 2012. "History of Slow Food." Accessed Nov. 28, 2012: http://www.slowfoodusa.org/index.php/about_us/details/manifesto/.

Smith, C. M., and H. E. Kurtz. 2003. "Community Gardens and Politics of Scale in New York City." *Geographical Review* 93 (2): 193–212. doi:10.1111/j.1931-0846.2003.tb00029.x.

Smith, Neil. 1998. "Giuliani Time: The Revanchist 1990s." *Social Text* 16 (4): 1–20. doi:10.2307/466878.

——. 1984. *Uneven Development: Nature, Capital and the Production of Space.* Oxford: Blackwell.

Solecki, William. 2011. "New York City as a Sustainable City." Columbia University Earth Institute Panel, Apr. 7, 2011.

Solid Waste Association of North America. 2002. *Successful Approaches to Recycling Urban Wood Waste.* Gen. Tech. Rep. FPL-GTR-133. Madison, WI: U.S. Department of Agriculture, Forest Service, Forest Products Laboratory.

Stein, Joshua. 2010. "What an Urban Farmer Looks Like," *New York Magazine,* Sept. 19, 2010. Accessed Dec. 15, 2010: http://nymag.com/restaurants/features/68297/.

Steinhauer, Jennifer. 2005. "Bloomberg's New Deputy Has a Velvet Fist" *New York Times,* Dec. 6, 2005. Accessed Dec. 8, 2011: http://www.nytimes.com/2005/12/06/nyregion/06patti.html?pagewanted=all.

Stone, C. N. 1993. "Urban Regimes and the Capacity to Govern: A Political Economy Approach." *Journal of Urban Affairs* 15: 1–28. doi:10.1111/j.1467-9906.1993.tb00300.x.

——. 1989. *Regime Politics: Governing Atlanta, 1946–1988.* Lawrence: University Press of Kansas.

Stone, Edie. 2009. "The Benefits of Community-managed Open Space: Community Gardening in New York City." In *Restorative Commons: Creating Health and Well-being through Urban Landscapes,* General Technical Report, edited by Lindsay Campbell and Anne Wiesen, 122–37. Newtown Square, PA: U.S. Department of Agriculture, Forest Service, Northern Research Station.

Strauss, A., and J. Corbin. 1998. *Basics of Qualitative Research: Techniques and Procedures for Developing Grounded Theory* 2nd ed. Thousand Oaks, CA: Sage Publications.

Stringer, Scott. 2014. "Audit Report on the Department of Parks and Recreation Tree Pruning Program." Office of the Comptroller. MD 13-107A. Aug. 15, 2014. Accessed Aug. 28, 2014: http://comptroller.nyc.gov/wp-content/uploads/documents/MD13_107A.pdf.

——. 2012. "Morning Plenary." Brooklyn Food Conference, May 12, 2012.

——. 2011. *Red Tape, Green Vegetables: A Plan to Improve New York City's Regulations for Community-Based Farmers Markets.* Apr. 2011. Accessed Nov. 28, 2012: http://www.libertycontrol.net/uploads/mbpo/RTGVReport.pdf.

——. 2010. *FoodNYC: A Blueprint for a Sustainable Food System.* Manhattan Borough President. Feb. 2010. Accessed Mar. 1, 2011: http://www.mbpo.org/uploads/policy_reports/mbp/FoodNYC.pdf.

——. 2009. *Food in the Public Interest.* Manhattan Borough President. Feb. 2009. Accessed Nov. 28, 2012: http://www.mbpo.org/uploads/policy_reports/FoodIn ThePublicInterest.pdf.

Strom, Stephanie. 2008. "Increase in Charitable Giving Dampened by Signs of Belt-Tightening." *New York Times,* June 23, 2008. Accessed June 4, 2013: http://www.nytimes.com/2008/06/23/us/23giving.html?_r=0.

Sullivan, Mercer. 1992. "Crime and the Social Fabric" In *Dual City: The Restructuring of New York,* edited by John Hull Mollenkopf and Manuel Castells, 225–44. New York: Russell Sage Foundation.

Susman, Tina. 2009. "A Tree Grows (and Dies) in Brooklyn." *Los Angeles Times,* Oct. 15, 2009. Accessed Nov. 12, 2012: http://articles.latimes.com/2009/oct/15/nation/na-trees15.

Susskind, L., L. Bacow, et al., eds. 1983. *Resolving Environmental Regulatory Disputes.* Cambridge, MA: Schenkman Books.

Susskind, L., and J. Cruikshank. 1987. *Breaking the Impasse: Consensual Approaches to Resolving Public Disputes.* New York, Basic Books.

Sustainable Flatbush. 2011. "Project Leaf Drop 2011: Compost Your Fall Leaves Locally." Accessed Nov. 21, 2012: http://sustainableflatbush.org/2011/11/07/project-leafdrop-2011-compost-your-fall-leaves-locally/.

Svendsen, Erika. 2013. "Storyline and Design: How Civic Stewardship Shapes Urban Design in New York City." In *Resilience in Ecology and Urban Design: Linking Theory and Practice for Sustainable Cities,* edited by S. T. A. Pickett, M. L. Cadenasso, and B. McGrath, 269–87. Dordrecht: Springer Netherlands.

——. 2010. *Civic Environmental Stewardship as a Form of Governance in New York City.* Ph.D. dissertation. New York: Columbia University.

——. 2009. "Cultivating Resilience: Urban Stewardship as a Means to Improving Health and Well-being." In *Restorative Commons: Creating Health and Well-being through Urban Landscapes*, General Technical Report, edited by Lindsay Campbell and Anne Wiesen, 58–87. Newtown Square, PA: U.S. Department of Agriculture, Forest Service, Northern Research Station.

Svendsen, Erika S., Lindsay K. Campbell, Dana R. Fisher, James J. T. Connolly, Michelle L. Johnson, Nancy F. Sonti, Dexter H. Locke, Lynne M. Westphal,

Cherie LeBlanc Fisher, J. Morgan Grove, Michele Romolini, Dale J. Blahna, and Kathleen L. Wolf. 2016. *Stewardship Mapping and Assessment Project: A Framework for Understanding Community-based Environmental Stewardship*. Gen. Tech. Rep. 156. Newtown Square, PA: U.S. Department of Agriculture, Forest Service, Northern Research Station.

Svendsen, Erika, and Jacqueline Lu. 2010. "MillionTreesNYC, Green Infrastructure, and Urban Ecology Symposium March 5–6, 2010." *Cities and the Environment* 3 (1): article 2. http://digitalcommons.lmu.edu/cate/vol3/iss1/2/.

Swyngedouw, E. 2005. "Governance Innovation and the Citizen: The Janus Face of Governance-beyond-the-state." *Urban Studies* 42 (11): 1991–2006. doi:10.1080/00420980500279869.

——. 2004. *Social Power and the Urbanization of Water: Flows of Power*. Oxford: Oxford University Press.

——. 1996. "The City as Hybrid: On Nature, Society, and Cyborg Urbanization." *Capitalism, Nature, Socialism* 7: 65–80. doi:10.1080/10455759609358679.

Swyngedouw, E., and N.C. Heynen. 2003. "Urban Political Ecology, Justice and the Politics of Scale." *Antipode* 35 (5): 898–918. doi:10.1111/j.1467-8330.2003.00364.x.

Tabb, William. 1982. *The Long Default*. New York: Monthly Review Press.

Taylor, A. F., and F. E. Kuo. 2009. "Children with Attention Deficits Concentrate Better after Walk in the Park." *Journal of Attention Disorders* 12 (5): 402–9. doi:10.1177/1087054708323000.

Taylor, A. F., F. E. Kuo, and W.C. Sullivan. 2001. "Coping with ADD: The Surprising Connection to Green Play Settings." *Environment and Behavior* 33 (1): 54–77. doi:10.1177/00139160121972864.

Taylor, B. 2009. "'Place' as Prepolitical Grounds of Democracy: An Appalachian Case Study in Class Conflict, Forest Politics, and Civic Networks." *American Behavioral Scientist* 52 (6): 826–45. doi:10.1177/0002764208327661.

Taylor, D.E. 2009. *The Environment and the People in American Cities, 1600s–1990s: Disorder, Inequality, and Social Change*. Durham, NC: Duke University Press.

Terry, Bryant. 2012. "Morning Plenary." Brooklyn Food Conference, May 12, 2012.

Tidball, Keith, Marianne Krasny, Erika Svendsen, Lindsay Campbell, and Kenneth Helphand. 2010. "Stewardship, Learning, and Memory in Disaster Resilience." *Environmental Education Research* 16 (5): 591–609. doi:10.1080/13504622.2010.505437.

Tranel, M., and L.B. Handlin. 2006. "Metromorphosis: Documenting Change." *Journal of Urban Affairs* 28 (2): 151–67. doi:10.1111/j.0735-2166.2006.00265.x.

Twiss, J., J. Dickinson, et al. 2003. "Community Gardens: Lessons Learned from California Healthy Cities and Communities." *American Journal of Public Health* 93 (9): 1435–38. doi:10.2105/AJPH.93.9.1435.

United Nations. 1992. *Agenda 21: The United Nations Program of Access from Rio*. Accessed May 10, 2010: http://www.un.org/esa/dsd/agenda21/index.shtml.

United States Census Bureau. 2013. State and County QuickFacts. Data derived from Population Estimates, American Community Survey, Census of Population and Housing, County Business Patterns, Economic Census, Survey of Business Owners, Building Permits, Census of Governments. Last Revised: Tuesday, 28-May-2013 09:50:40 EDT. Accessed June 6, 2013: http://quickfacts.census.gov/qfd/states/36/3651000.html.

——. 2010. Census of Population, Public Law 94-171 Redistricting Data File. Updated every 10 years. Accessed Sept. 28, 2012: http://factfinder2.census.gov.

United States Department of Agriculture. 2012. "Grants, Loans, and Support." Know Your Farmer, Know Your Food. Accessed Nov. 28, 2012: http://www.usda.gov/wps/portal/usda/usdahome?navid=KYF_GRANTS).

U.S. Forest Service. 2013. "New York City Urban Field Station" webpage. Accessed Aug. 2, 2013: http://www.nrs.fs.fed.us/nyc/.

——. 2007. *Stewardship Mapping and Assessment Project* [computer file]. New York: U.S. Department of Agriculture, Forest Service.

Van Ooyen, Marcel. 2012. "The Future of NYC Food Policy." Workshop Panel, Brooklyn Food Conference, May 12, 2012.

Vayda, Andrew P., and Bradley Walters. 1999. "Against Political Ecology." *Human Ecology* 27 (1): 167–79. doi:10.1023/A:1018713502547.

Von Hassel, M. 2002. *The Struggle for Eden: Community Gardens in New York City.* Westport, CT: Bergin and Garvey.

Waitt, Gordan. 2005. "Doing Discourse Analysis." In *Qualitative Methods in Human Geography* 2nd ed., edited by Iai Hay, 163–91. Oxford: Oxford University Press.

Wakefield, S., F. Yeudall, C. Taron, J. Reynolds, and A. Skinner. 2007. "Growing Urban Health: Community Gardening in South-East Toronto." *Health Promotion International* 22 (2): 92–101. doi:10.1093/heapro/dam001.

Walker, G. 2003. "Reconsidering Regional Political Ecologies? Toward a Political Ecology of the Rural American West." *Progress in Human Geography* 27 (1): 7–24. doi: 10.1191/0309132503ph410oa.

Wapner, Paul. 1995. "Politics Beyond the State: Environmental Activism and World Civic Politics." *World Politics* 47 (3): 311–40. doi:10.1017/S0043887100016415.

Ward, Kathleen T., and Gary R. Johnson. 2007. "Geospatial Methods Provide Timely and Comprehensive Urban Forest Information." *Urban Forestry & Urban Greening* 6 (1): 15–22. doi:10.1016/j.ufug.2006.11.002.

Wasserman, Stanley, and Katherine Faust. 1994. *Social Network Analysis: Methods and Applications*. Cambridge, UK: Cambridge University Press.

Watt, Fiona. 2007. "PlaNYTree." NYC Department of Parks and Recreation internal memo, June 26, 2007.

Weber, Christopher L., and H. Scott Matthews. 2008. "Food-Miles and the Relative Climate Impacts of Food Choices in the United States." *Environ. Sci. Technol.* 42 (10): 3508–13. doi:10.1021/es702969f.

Wellman, B. 1979. "The Community Question: The Intimate Networks of East Yorkers." *American Journal of Sociology* 85 (5): 1201–31.

Wells, N. M. 2003. "Nearby Nature: A Buffer of Life Stress among Rural Children." *Environment and Behavior* 35 (3): 311–30. doi:10.1177/0013916503035003001.

Wells, Pete. 2010. "Cooking with Dexter: Not in My Backyard." *New York Times,* Oct. 20, 2010. Accessed Oct. 31, 2010: http://www.nytimes.com/2010/10/24/magazine/24food-t-000.html?ref=urban_agriculture.

Whatmore, S. 2006. "Materialist Returns: Practicing Cultural Geography in and for a More-than-human World." *Cultural Geographies* 13: 600–609. doi:10.1191/1474474006cgj377oa.

Whatmore, S., and L. Thorne. 2000. "Elephants on the Move: Spatial Formations of Wildlife Exchange." *Environment and Planning-D* 18 (2): 185–203. doi:10.1068/d210t.

While, Aidan, Andrew Jonas, and David Gibbs. 2004. "The Environment and the Entrepreneurial City: Searching for the Urban 'Sustainability Fix' in Manchester and Leeds." *International Journal of Urban and Regional Research* 28 (3): 549–69. doi:10.1111/j.0309-1317.2004.00535.x.

Whitsett, Ross. 2006. "Urban Mass: A Look at Co-Op City." *The Cooperator: The Coop and Condo Monthly*, Dec. 2006. Accessed Nov. 8, 2012: http://cooperator.com/articles/1354/1/Urban-Mass/Page1.html.

Williams, Jumane. 2012. "How the NYC Government Can Impact Our Food System." Panel Session, Brooklyn Food Conference, May 12, 2012.

Wilson, James Q., and George L. Kelling. 1982. "The Police and Public Safety: Broken Windows." *Atlantic Monthly*, Mar. 1982.

Wolch, J. 1990. *The Shadow State: Government and Voluntary Sector in Transition*. New York: The Foundation Center.

Wolf, K. L. 2005. "Business District Streetscapes, Trees and Consumer Response." *Journal of Forestry* 103 (8): 396–400. http://www.treesearch.fs.fed.us/pubs/34952.

——. 2004. "Trees and Business District Preferences: A Case Study of Athens, Georgia US." *Journal of Arboriculture* 30 (6): 336–46. http://www.treesearch.fs.fed.us/pubs/34955.

——. 2003. "Public Response to the Urban Forest in Inner-City Business Districts." *Journal of Arboriculture* 29 (3): 117–26. http://www.treesearch.fs.fed.us/pubs/34957.

Wu, Felicia, and Tim K. Takaro. 2007. "Childhood Asthma and Environmental Interventions." *Environmental Health Perspectives* 115 (6): 971–75. doi:10.1289/ehp.8989.

Ye, Jenny, and Matthew Schuerman. 2014. "If a Tree Dies in the City, Will Anyone Notice?" WNYC. Accessed Aug. 28, 2014: http://www.wnyc.org/story/if-tree-dies-city-will-anyone-notice/.

Young, Iris Marion. 2000. *Inclusion and Democracy*. New York: Oxford University Press.

Index

A *t* or *f* following a page number denotes a table or figure

www.ingramcontent.com/pod-product-compliance
Ingram Content Group UK Ltd.
Pitfield, Milton Keynes, MK11 3LW, UK
UKHW041236290525
459016UK00013B/459/J

9 781501 707506